Summary

In the past, the oil and gas industry considered gas locked in tight, impermeable shale uneconomical to produce. However, advances in directional well drilling and reservoir stimulation have dramatically increased gas production from unconventional shales. The United States Geological Survey estimates that 200 trillion cubic feet of natural gas may be technically recoverable from these shales. Recent high natural gas prices have also stimulated interest in developing gas shales. Although natural gas prices fell dramatically in 2009, there is an expectation that the demand for natural gas will increase. Developing these shales comes with some controversy, though.

The hydraulic fracturing treatments used to stimulate gas production from shale have stirred environmental concerns over excessive water consumption, drinking water well contamination, and surface water contamination from both drilling activities and fracturing fluid disposal.

The saline "flowback" water pumped back to the surface after the fracturing process poses a significant environmental management challenge in the Marcellus region. The flowback's high content of total dissolved solids (TDS) and other contaminants must be disposed of or adequately treated before discharged to surface waters. The federal Clean Water Act and state laws regulate the discharge of this flowback water and other drilling wastewater to surface waters, while the Safe Drinking Water Act (SDWA) regulates deep well injection of such wastewater. Hydraulically fractured wells are also subject to various state regulations. Historically, the EPA has not regulated hydraulic fracturing, and the 2005 Energy Policy Act exempted hydraulic fracturing from SDWA regulation. Recently introduced bills would make hydraulic fracturing subject to regulation under SDWA, while another bill would affirm the current regulatory exemption.

Gas shale development takes place on both private and state-owned lands. Royalty rates paid to state and private landowners for shale gas leases range from 12½% to 20%. The four states (New York, Pennsylvania, Texas, and West Virginia) discussed in this report have shown significant increases in the amounts paid as signing bonuses and increases in royalty rates. Although federal lands also overlie gas shale resources, the leasing restrictions and the low resource-potential may diminish development prospects on some federal lands. The practice of severing mineral rights from surface ownership is not unique to the gas shale development. Mineral owners retain the right to access surface property to develop their holdings. Some landowners, however, may not have realized the intrusion that could result from mineral development on their property.

Although a gas-transmission pipeline-network is in place to supply the northeast United States, gas producers would need to construct an extensive network of gathering pipelines and supporting infrastructure to move the gas from the well fields to the transmission pipelines, as is the case for developing any new well field.

Contents

Figures

Tables

Appendixes

Contacts

Background

Until recently, many oil and gas companies considered natural gas locked in tight, impermeable shale uneconomical to produce. Advanced drilling and reservoir stimulation methods have dramatically increased the gas production from "unconventional shales." The Barnett Shale formation in Texas has experienced the most rapid development. The Marcellus Shale formation of the Appalachian basin, in the northeastern United States, potentially represents the largest unconventional gas resource in the United States. Other shale formations, such as the Haynesville shale, straddling Texas and Louisiana, have also attracted interest, as have some formations in Canada. The resource potential of the shales has significantly increased the natural gas reserve estimates in the United States.[1]

Recent high natural gas prices had stimulated interest in gas shales, although prices fell below $4.00 per million Btu (mmBtu) in the summer of 2009.[2] Despite the currently low natural gas prices, there is an expectation that the demand for natural gas will increase and that unconventional gas shales will play a major role in meeting the increased demand for natural gas.

The shale's development is a subject of controversy, however. The potential economic benefits from both the drilling activities and the lease and royalty payments compete with the public's concern for environmentally safe drilling practices and protection of groundwater and surface water.

Directional drilling and "hydraulic fracturing" are instrumental in exploiting this resource. Although oil and gas developers have applied these technologies in conventional oil fields for some time, they have only recently begun applying them to unconventional gas shales. As with oil and gas production generally, gas shale development is primarily subject to state law and regulation, although provisions of two relevant federal laws—the Safe Drinking Water Act (SDWA) and the Clean Water Act (CWA)—also apply. Regulation of well construction differs by state, and federal law currently exempts from regulation the underground injection of fluids for hydraulic fracturing purposes. Two recently introduced bills would subject hydraulic fracturing to regulation under the Safe Drinking Water Act, while a third bill would affirm the current SDWA exemption. There has been an increase in reports of public concerns that hydraulic fracturing poses a potential risk to water wells and thus domestic drinking water supply. The concern is that the chemicals used pose a potential risk to groundwater quality, and the fracturing may damage aquifers. Critics maintain that the large quantities of water that hydraulic fracturing consumes may tax local and regional water supplies and that disposing the "flowback" extracted from the shale after fracturing may affect surface water and groundwater quality. Understanding the technical principles of well drilling, well construction, and stimulation methods are important to assessing the concerns that have arisen over the shale's development.

---

[1] "Reserves are those quantities of petroleum, which, by analysis of geoscience and engineering data, can be estimated with reasonable certainty to be commercially recoverable ... " in: Society of Petroleum Engineers: *Petroleum Resources Management System,* 2007, p. 3, http://www.spe.org/spe-app/spe/industry/reserves/prms.htm.

[2] See Energy Information Administration, *Natural Gas page*, http://www.eia.doe.gov/oil_gas/natural_gas/info_glance/ natural_gas.html.

Gas shale development takes place on both private and state lands. Royalty rates, rents, and signing bonuses paid to state and private landowners vary. In Texas, where similar gas shale development takes place, the rates are generally higher. Although federal lands overlie some portions of the Marcellus, restrictions to drilling and low resource potential may make them less desirable to develop. In the case of split-estate lands, which sever mineral rights from surface ownership, the mineral-owner retains the right to access the land for development. Landowners may not be aware of the mineral-owners' access rights, and are not due compensation for access. Furthermore, they may have to contend with long-term easements for the gathering pipelines that connect the wells to natural gas transmission lines. The states do regulate reclamation of the drilling sites to varying degrees.

This report does not discuss all unconventional gas shales. Rather, this report limits discussion to the Barnett and Marcellus Shale formations, which are most representative of the resource. Both formations also serve to illustrate the technical and policy issues that are most likely common to developing all gas shales.

Unconventional Gas Shale Resources in the United States

Unconventional gas shales are fine grained, organic rich, sedimentary rocks. The shales are both the source of and the reservoir for natural gas, unlike conventional petroleum reservoirs. In the shales, gas occupies pore spaces, and organic matter adsorbs gas on its surface. The Society of Petroleum Engineers describes "unconventional resources" as petroleum accumulations that are pervasive throughout a large area and that are not significantly affected by hydrodynamic influences (they are also called "continuous-type deposits"). In contrast, conventional petroleum and natural gas occur in porous sandstone and carbonate reservoirs. Under hydrodynamic pressure exerted by water, the petroleum migrated upward from its organic source until an impermeable cap-rock (such as shale) trapped it in the reservoir rock. The "gas-cap" that accumulated over the petroleum has been the source of most produced natural gas.

Though the shales may be as porous as other sedimentary reservoir rocks, their extremely small pore sizes make them relatively impermeable to gas flow, unless natural or artificial fractures occur. Major gas shale basins exist throughout the lower-48 United States. There are at least 21 shale basins in more than 20 states (see **Figure 1**).[3]

Based on a recent assessment of natural gas resources, the United States has a base of 1,836 trillion cubic-feet (tcf).[4] Shale gas made up an estimated one-third of the resource base, roughly 616 tcf. Stated in other terms, shale gas represents the equivalent of approximately 102 billion barrels of crude oil.

---

[3] *North American Natural Gas Supply Assessment*, Prepared for Clean Skies Foundation, Navigant Consulting, July 4, 2008.

[4] Colorado School of Mines, *Potential Gas Committee reports unprecedented increase in magnitude of U.S. natural gas resource base*, June 18, 2009, http://www.mines.edu/Potential-Gas-Committee-reports-unprecedented-increase-in-magnitude-of-U.S.-natural-gas-resource-base.

Figure 1. Major Shale Basins in the Conterminous United States

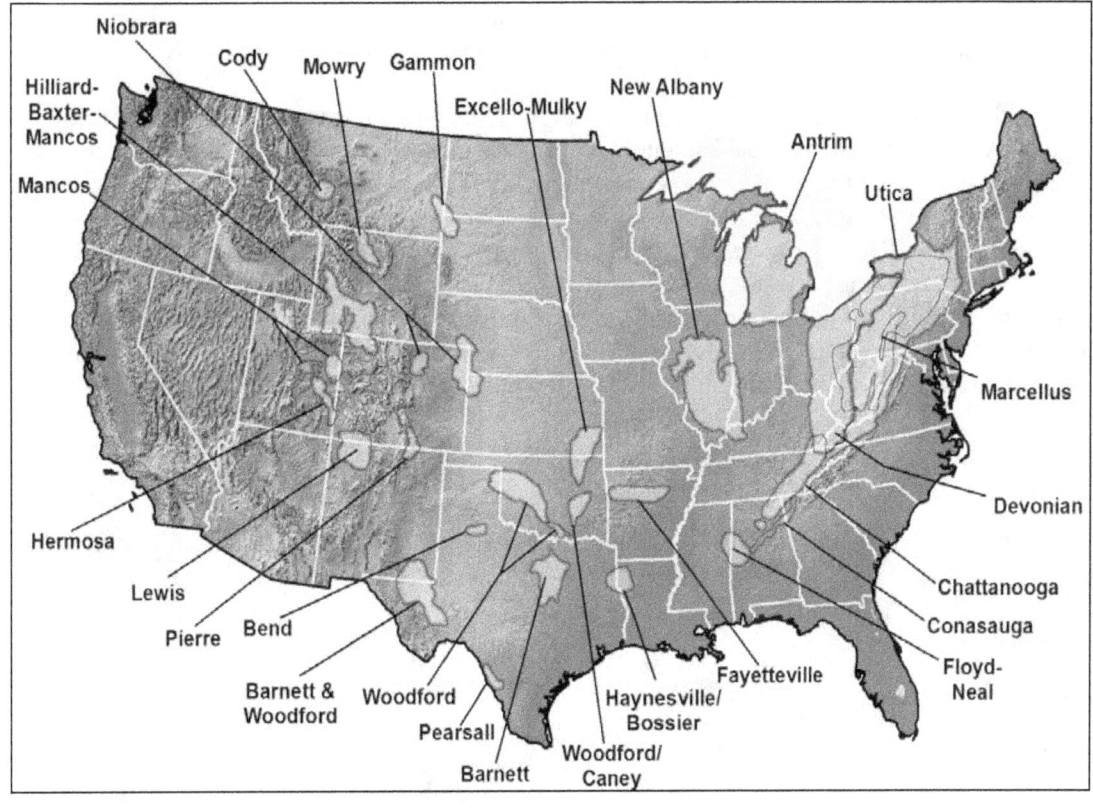

Source: U.S. DOE Office of Fossil Energy, *Modern Shale Gas Development in the United States: A Primer.*

Natural gas production in the United States had been declining until recently. After peaking at nearly 24.2 tcf in 2000, production declined to 23.5 tcf by 2005.[5] The decline had reversed by 2008 when production rose to over 26 tcf. (Gulf of Mexico offshore gas production continues to decline, however, and the 2.3 tcf produced in 2008 was only 47% of the 2000 Gulf production level.)

Since 1990, U.S. gas production increased in large part due to steadily increasing production of unconventional sources. The U.S. Geological Survey (USGS) estimates that about 200 tcf of natural gas may be technically recoverable from the shale with a recovery factor of 10% to 15%.[6] These sources include shale gas, coal-bed methane, and tight sands. Unconventional gas production increased 65% from 1998 (5.4 tcf/year) to 2007 (8.9 tcf/year).[7]

---

[5] U.S. DOE Energy Information Administration, *Natural Gas Navigator*, Natural Gas Summary, June 29, 2009, http://tonto.eia.doe.gov/dnav/ng/ng_sum_lsum_a_EPG0_FGW_mmcf_a.htm.

[6] U.S. DOE Energy Information Administration, I*s U.S. natural gas production increasing?*, June 11, 2008, http://tonto.eia.doe.gov/energy_in_brief/natural_gas_production.cfm.

[7] Navigant Consulting, *Unconventional Gas Supply*, TransCanada Pipelines Customer Meeting, San Francisco, CA,

(continued...)

Natural gas production in the "big 7" shale plays (Antrim, Barnett, Devonian, Fayetteville, Woodford, Haynesville, Marcellus) could reach an estimated 27 to 39 bcf/day within 10 to 15 years.[8] Development, however, has been uneven in the Marcellus. Although production data are incomplete at best, reports indicate that Pennsylvania leads the region in producing wells, with West Virginia and Ohio trailing. Navigant expects that the Marcellus basin will be the next significant gas shale play. It appears potentially larger than the other basins already developed.

In 2006, the Bank of America estimated break-even costs for shale gas production to be within a range of $4.20 to $11.50 per thousand cubic feet (mcf).[9] The estimated median break-even cost was $6.64 per mcf. Leading shale gas producers include Chesapeake Energy Corporation, Devon Energy, XTO Energy, Inc., Southwestern Energy, Newfield Exploration Company, and Encana. In late 2008, StatoilHydro, the second largest natural gas supplier to Europe, and Chesapeake Energy Corporation, the largest U.S. natural gas producer, announced a joint agreement to explore unconventional gas opportunities. Under these agreements StatoilHydro will initially acquire a 32.5% interest in Chesapeake's Marcellus Shale gas acreage covering 1.8 million net acres (StatoilHydro's share equals approximately 0.6 million net acres of the leasehold).[10]

Barnett Shale Formation

The Barnett Shale formation is Mississippian-age black shale that has a high organic content.[11] It underlies 5,000 square miles of the Dallas/Fort Worth area of Texas (primarily in the Fort Worth Basin) to depths of 6,500 to 8,500 feet. (See **Figure 2**.)

Natural Gas Resource Potential

The Barnett Shale play is reportedly the most active natural gas play in the United States with as many as 173 drilling rigs at work in the past year. The USGS estimated that as much as 26.7 tcf of natural gas could be present in continuous accumulations as nonassociated gas trapped in strata of two of the three Barnett Shale Assessment Units (AU)—the Greater Newark East Frac-Barrier Continuous Barnett Shale Gas AU and the Extended Continuous Barnett Shale Gas AU.[12] Collectively, the units as comprise the Bend Arch–Fort Worth Basin.

(...continued)

August 26, 2008, http://www.navigantconsulting.com/downloads/knowledge_center/ Unconventional_Natural_Gas_RWelch_Aug_08.pdf.

[8] See Navigant Consulting.

[9] See Navigant Consulting.

[10] StatoilHydro, "Forms strategic alliance with major US gas player," press release, November 11, 2008, http://www.statoilhydro.com/en/NewsAndMedia/News/2008/Pages/Chesapeake.aspx.

[11] Mississippian-aged shale dates back to the Paleozoic era of around 345 to 320 million years ago.

[12] Richard M. Pollastro (Task Leader), Ronald J. Hill, and Thomas A. Albrandt, et al., Assessment of Undiscovered Oil and Gas Resources of the Bend Arch-Fort Worth Basin Province of North-Central Texas and Southwestern Oklahoma, 2003, U.S. Geological Survey, Fact Sheet 2004-3022, March 2004. http://pubs.usgs.gov/fs/2004/3022/

The area has produced oil and gas since the early 1900s, but mostly from conventional reservoirs.[13] There are over 8,000 wells producing gas from the Barnett formation.[14] Gas production increased from 94 million cubic feet per day (mmcf/day) in 1998 to over 3 billion cubic feet/day (bcf/day) in 2007; an increase of over 3,000%.

Figure 2. Bend Arch-Fort Worth Basin Area

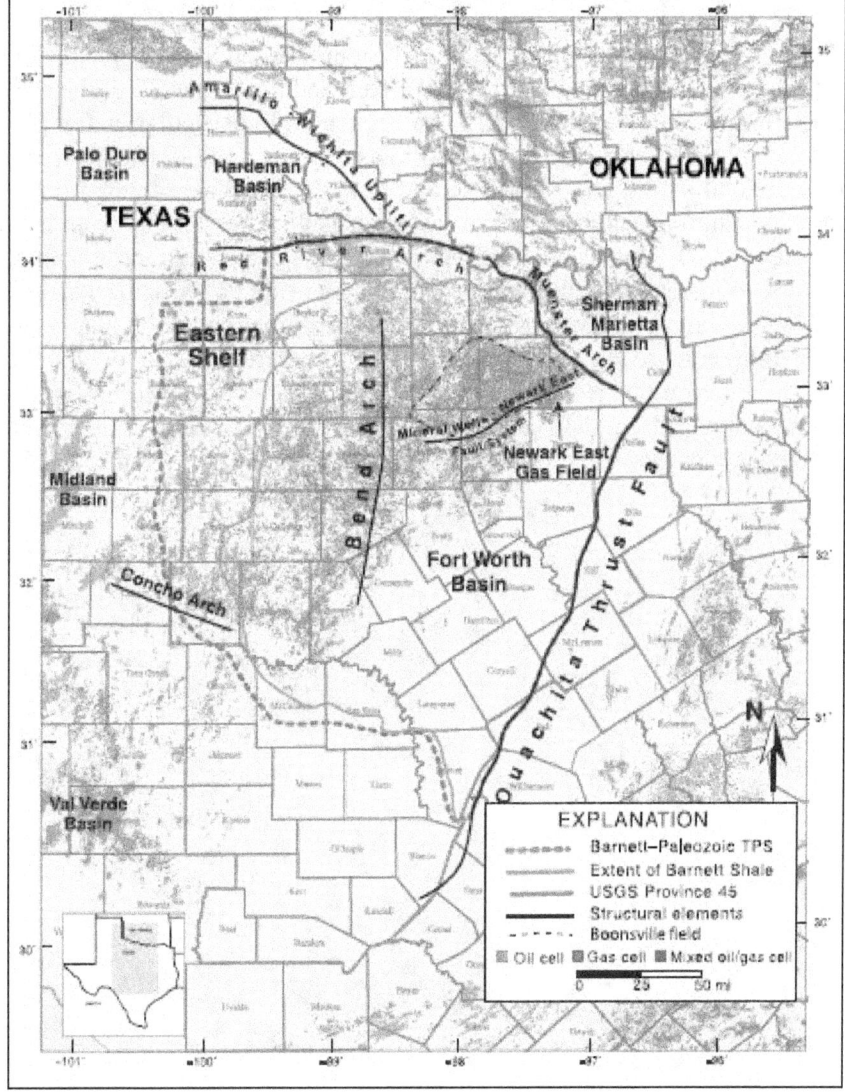

Source: Pollastro et al, *Geologic Framework of the Mississippian Barnett Shale, Barnett-Paleozoic Total Petroleum System, Bend Arch-Fort Worth Basin*, Texas, AAPG Bulletin, V. 91, No. 4 (April 2007), pp 405-436.

---

[13] Richard M. Pollastro, Daniel M. Jarvie, and Ronald J. Hill, et al., "Geologic Framework of the Mississippian Barnett Shale, Barnett-Paleozoic total petroleum system, Bend arch-Fort Worth Basin, Texas," *American Association of Petroleum Geologists*, vol. 91, no. 4 (April 2007), pp. 405-436.

[14] Personal communication with Eric Potter, Jackson School of Geosciences, University of Texas, October 2008.

The Newark East field of the Barnett has been the largest gas-producing field in Texas since 2000, and ranked as the second highest U.S. gas-producing field in 2005. Gas production rose from less than 11 bcf of natural gas in 1993 to about 480 bcf by 2005.[15] From January 1993 to January 2006, cumulative gas production measured about 1.8 tcf. The Newark East field proven gas reserve estimates range from between 2.5 and 3.0 tcf.

Southwest Regional Gas Supply and Demand

Several major pipeline corridors transport natural gas in the United States. Five major pipeline routes extend from the producing areas in the Southwest, and more than 20 of the major interstate pipelines originate in the Southwest Region.[16] (See **Figure 3**.) In particular, major pipeline networks, constructed after World War II, transport gas from the Gulf of Mexico region to the Northeast, from the Southwest to the Northeast, and from the Southwest to the Midwest.[17] Current pipelines have the capacity to transport as much as 45.2 bcf per day from the region: 62% to the Southeast Region, 20% to the Central Region, 13% to the Western Region, and the rest to Mexico. In response to increased natural gas production and supply, particularly from the Barnett Shale, the Southwest region recently expanded its pipeline infrastructure on a large-scale. In 2008, 30 new pipelines comprising 1,382 miles reached completion; nearly double the previous year's. Thirteen of the new pipelines related to or expanded the northeast Texas area to new development of gas supplies from the Barnett, Woodford, or Fayetteville Shale formations. The remaining pipelines support increased liquefied natural gas (LNG) imports through Texas marine terminals.

---

[15] Pollastro, et al., *Geologic framework of the Mississippian Barnett Shale.*

[16] Energy Information Administration, Office of Oil and Gas, Natural Gas Division , *About U.S. Natural Gas Pipelines*, Major Natural Gas Transportation Corridors, http://www.eia.doe.gov/pub/oil_gas/natural_gas/analysis_publications/ngpipeline/transcorr.html.

[17] Damien Gaula, *Expansion of the U.S. Natural Gas Pipeline Network: Additions in 2008 and Projects through 2011*, Energy Information Administration, Office of Oil and Gas, September 2009, http://www.eia.doe.gov/pub/oil_gas/natural_gas/feature_articles/2009/pipelinenetwork/pipelinenetwork.pdf.

Figure 3. Major Natural Gas Transportation Corridors in the Conterminous United States, 2008

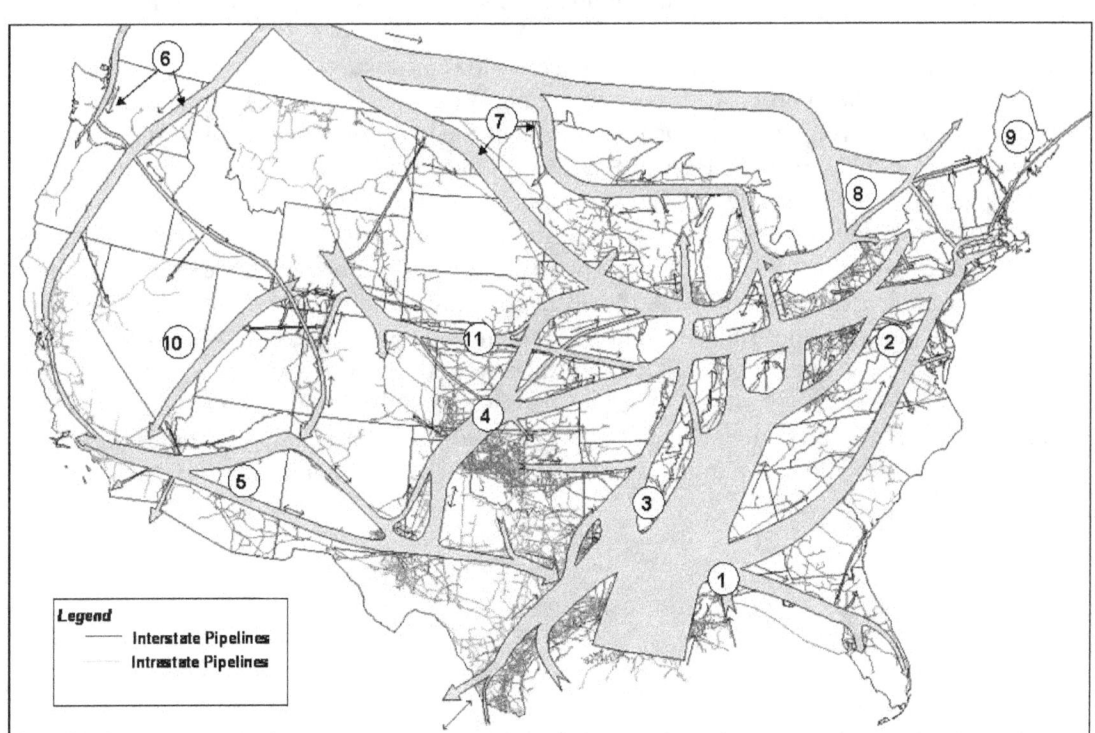

Source: EIA, Gas Tran Gas Transportation Information System http://www.eia.doe.gov/pub/oil_gas/natural_gas/analysis_publications/ngpipeline/transcorr_map.html.

Notes: major pipeline networks originating from the Southwest are: (1) Southwest-Southeast, (2) Southwest-Northeast, (3) Southwest-Midwest, (4) Southwest Panhandle-Midwest, and (5) Southwest-Western. The EIA has determined that this informational map does not raise security concerns, based on the application of the Federal Geographic Data Committee's *Guidelines for Providing Appropriate Access to Geospatial Data in Response to Security Concerns*.

Groundwater Resource Issues

As gas development in the Barnett Shale increased over the past decade, water use also increased. One study estimated that water used for Barnett Shale gas development increased from approximately 700 acre-feet (AF) in 2000 to more than 7,000 AF in 2005.[18] Barnett Shale development uses both surface water and groundwater resources, primarily for hydraulically fracturing vertical and horizontal wells. Depending on the well type, from 1.2 to 3.5 million gallons (4 to 11 AF) of water may be consumed in hydraulically fracturing a gas well. Of the approximately 7,000 AF used in 2005, about 60% came from groundwater in the Trinity and

---

[18] James Bene et al., *Northern Trinity/Woodbine Groundwater Availability Model: Assessment of Groundwater Use in the Northern Trinity Aquifer Due to Urban Growth and Barnett Shale Development*, R.W. Harden & Associates, prepared for the Texas Water Development Board, Austin, TX, January 2007, p. 2-21, at http://www.twdb.state.tx.us/RWPG/rpgm_rpts/0604830613_BarnetShale.pdf. Hereafter referred to as R.W. Harden & Associates, 2007.

Woodbine aquifers in north central Texas.[19] The deeper Trinity aquifer (underlying 61 counties) is much more extensive than the shallower Woodbine aquifer (underlying 17 counties).[20] The state of Texas also considers Trinity a major aquifer compared to the minor Woodbine.[21]

The Trinity Aquifer extends from south-central Texas to southeastern Oklahoma in a 550 mile-long arc-like band.[22] (See **Figure 4**.) It is an important water source for many communities, particularly rural, in north-central Texas where Barnett Shale development is most intense. Groundwater use varies across the Barnett Shale development area; for example, groundwater provides approximately 85% of the total water supply in Cooke County but only 1% for Dallas County.[23] Extensive development of the Trinity Aquifer in the Dallas-Ft. Worth metropolitan areas had caused groundwater levels to drop over 500 feet in some areas. However, when their populations began increasing after the 1970s, local communities abandoned many public supply wells in favor of surface water supplies. This resulted in a recovery of groundwater levels to some extent.[24] For many rural areas, however, groundwater from the Trinity Aquifer remains the sole water source. The Woodbine Aquifer, which is shallower than the Trinity, offers another groundwater supply for the region. The lowest of its three water-bearing layers yields the most water. Though the Woodbine is a minor aquifer, it could be locally important to Barnett Shale development given its proximity.

---

[19] R.W. Harden & Associates, 2007, p. 2.

[20] Texas Water Development Board, *2007 State Water Plan, Chapter 7, Groundwater Resources*.

[21] In Texas, a minor aquifer is defined as one that supplies large quantities of water in small areas or small quantities of water in large areas; a major aquifer is generally defined as supplying large quantities of water over a large areas of the state. See John B. Ashworth and Janie Hopkins, *Major and Minor Aquifers of Texas*, Texas Water Development Board, Report 345, November 1995, p. 1. The state of Texas recognizes 7 major and 21 minor aquifers, see Texas Water Development Board, *2007 State Water Plan, Chapter 7, Groundwater Resources*, at http://www.twdb.state.tx.us/wrpi/swp/swp.htm.

[22] Paul D. Ryder, U.S. Geological Survey, *Ground Water Atlas of the United States*, Oklahoma, Texas; HA 730-E, 1996, at http://pubs.usgs.gov/ha/ha730/ch_e/index.html.

[23] R.W. Harden & Associates, 2007, p. 2.

[24] John B. Ashworth and Janie Hopkins, *Major and Minor Aquifers of Texas: Trinity Aquifer*, Texas Water Development Board, Report 345, November 1995.

Figure 4. Aquifers in the Bend Arch-Fort Worth Basin Area

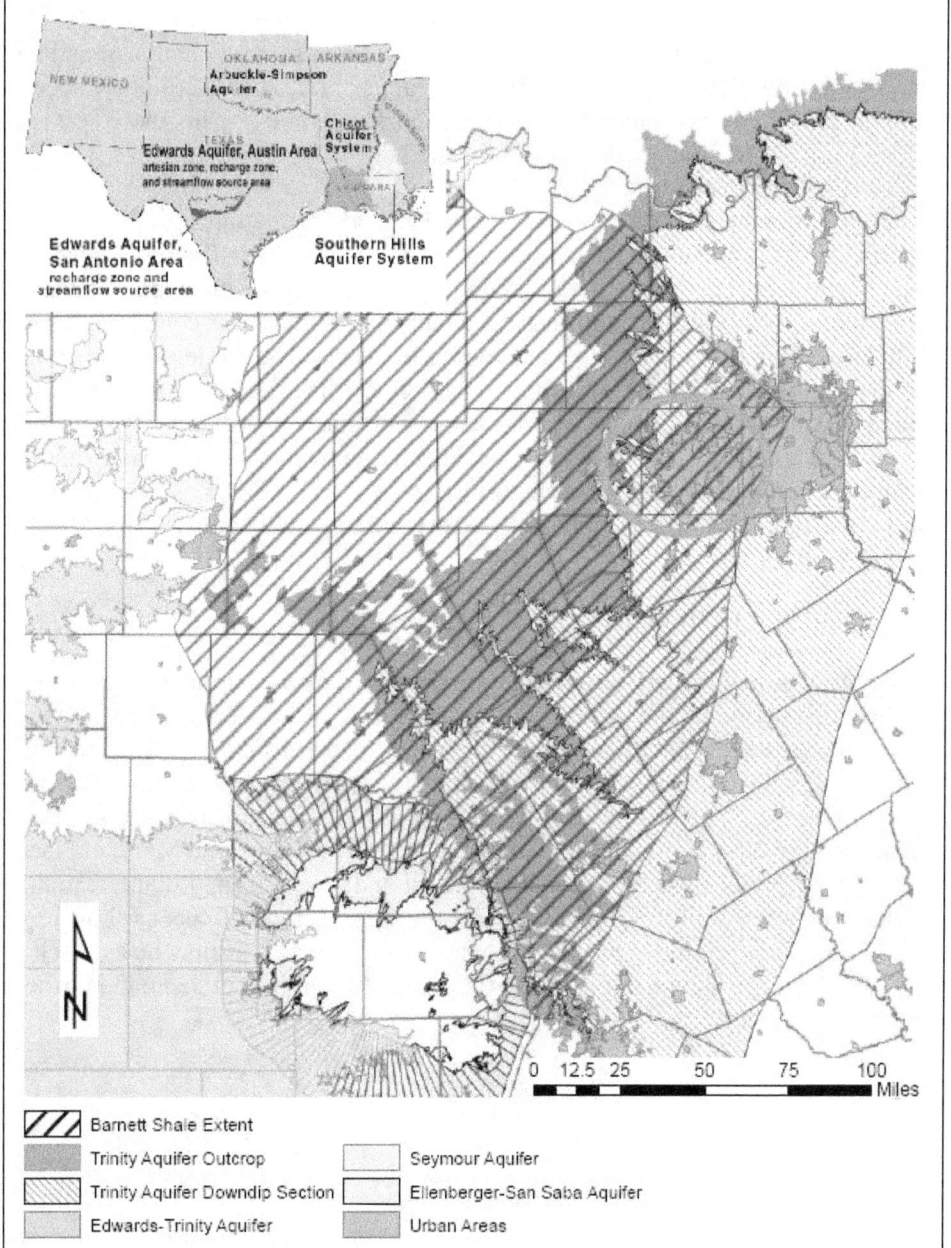

Source: James Bene et al., Northern Trinity/Woodbine Groundwater Availability Model: Assessment of Groundwater Use in the Northern Trinity Aquifer Due to Urban Growth and Barnett Shale Development, R.W. Harden & Associates. p. 2-5 Corner Inset Map: U.S. EPA, *Sole Source Aquifer Protection Program*, Regions VI (modified by CRS) .

Notes: Green circled area represents the so-called "core area" where most of the Barnett Shale gas development has taken place.

A study that modeled groundwater use in the area concluded that about 3% of the use was associated with the Barnett Shale's development in 2005.[25] Depending on the pace of shale development, water use (which is essential to hydraulic fracturing) could vary widely.[26] According to the study, low natural gas prices would slow interest in development and thus slightly decrease groundwater use by 2025. High natural gas prices could stimulate accelerated development and thus increase groundwater demand for development to about 10,000 to 25,000 AF by 2025, or from 3% in 2005 to 7% to 13% by 2025.

Increasing shale development could compete with other users for the same groundwater resources, particularly in rural areas where groundwater is a significant fraction of water supplies. It is uncertain how urban areas may react to groundwater well drilling in support of shale development (that is, for hydraulic fracturing), even though groundwater comprises a smaller fraction of the water supply compared to some more rural areas. The modeling study pointed out, as well, that the Trinity and Woodbine aquifers underlie only the eastern portion of the Barnett Shale; no major or minor aquifers underlie the western portions.[27] Development of western portions raises the possibility of transporting groundwater pumped from the Trinity and Woodbine aquifers in the east to the western portion of the Barnett Shale, which could raise estimates of potential groundwater use above those presented in the model results.

Marcellus Shale Formation

The Marcellus Shale is sedimentary rock formation deposited over 350 million years ago during the middle-Devonian period on the geologic timescale. Geologic strata deposited in the Appalachian basin during this period are likely to produce more gas than oil. Regional oil production is associated with Pennsylvanian age strata (of the later Carboniferous period). Most of this black, organic-rich shale lies beneath much of West Virginia, western and northeastern Pennsylvania, southern New York, eastern Ohio, and parts of Virginia and Maryland (see **Figure 5**). It is an estimated 95,000 square miles in areal extent and ranges from 4,000 feet to 8,500 feet in depth.[28] The shale's non-uniform thickness varies from 50 feet to 250 feet as shown by the isopach map of **Figure 5**.[29] Some reports indicate that shale's thickness may be as much as 900 feet in places, however. As shown in **Figure 6**, the Marcellus plunges in depth the further north it goes along the cross section.

---

[25] James Bene et al., *Northern Trinity/Woodbine Groundwater Availability Model: Assessment of Groundwater Use in the Northern Trinity Aquifer Due to Urban Growth and Barnett Shale Development*, R.W. Harden & Associates, p.2.

[26] Other factors could include the thickness and maturity of the shale, the ability to drill horizontally or vertically, the number of well completions per year, availability of water, and regulatory factors, among others.

[27] James Bene et al., *Northern Trinity/Woodbine Groundwater Availability Model: Assessment of Groundwater Use in the Northern Trinity Aquifer Due to Urban Growth and Barnett Shale Development*, R.W. Harden & Associates, p.15.

[28] U.S. Department of Energy Office of Fossil Energy and National Technology Laboratory, *Modern Shale Gas Development in the United States: A Primer*, DE-FG26-04NT15455, April 2009, http://fossil.energy.gov/programs/oilgas/publications/naturalgas_general/Shale_Gas_Primer_2009.pdf.

[29] An isopach is a continuous line connecting points of equal thickness.

Figure 5. Marcellus Shale Formation Thickness

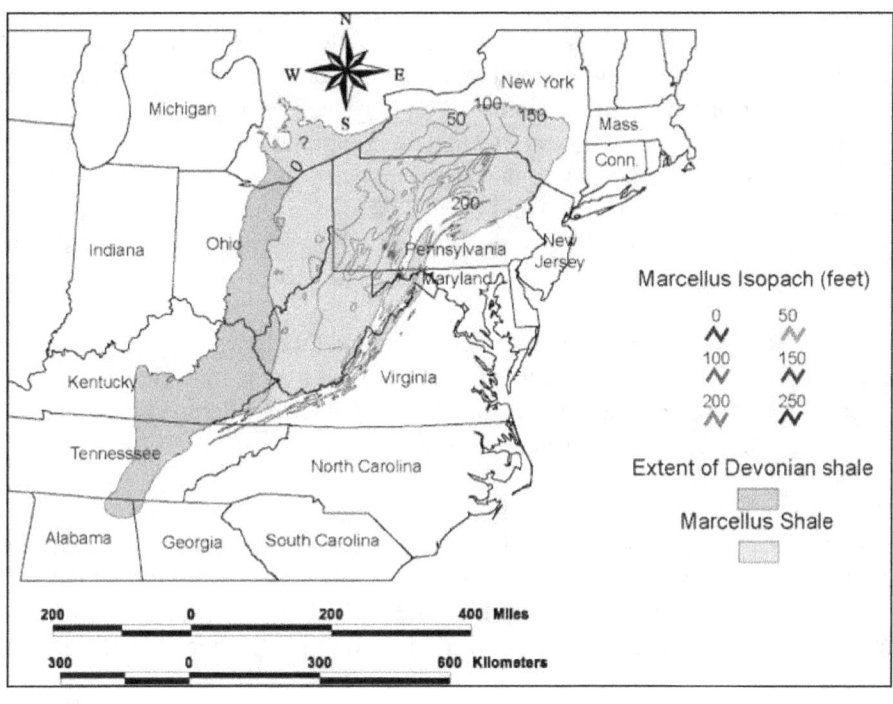

Source: U.S. Geological Survey, *Assessment of Undiscovered Natural Gas Resources in Devonian Black Shales, Appalachian Basin, Eastern U.S.A.,* Open-File Report 200-1268, 2005.

Figure 6. Devonian Shale Cross Section

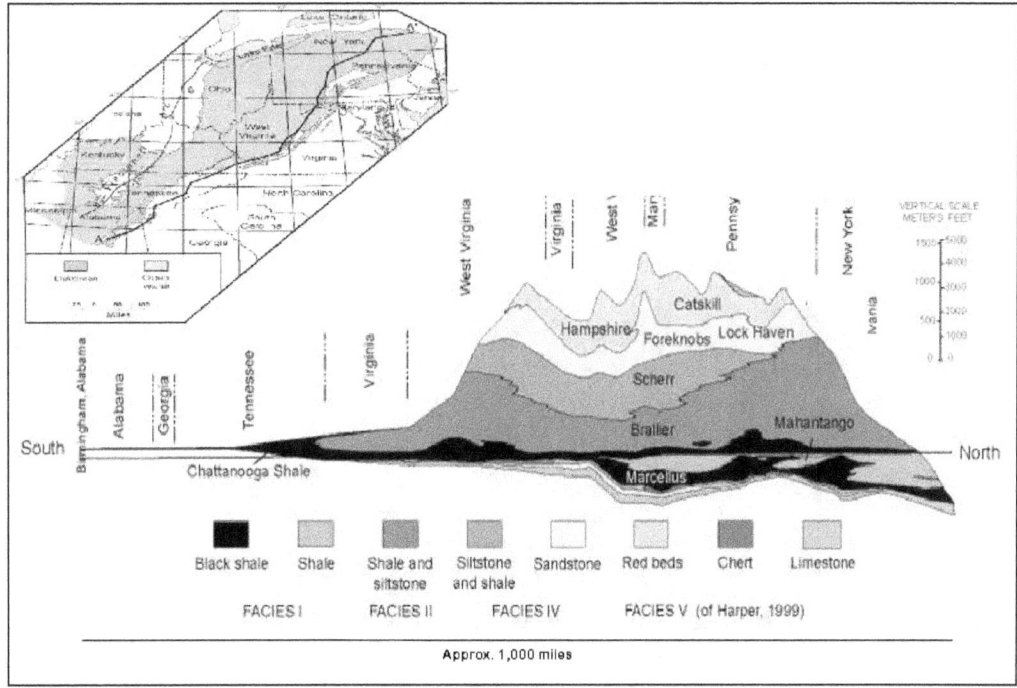

Source: Compiled from USGS Open File Report 200-1268, 2005.

Notes: Cross section from Alabama to New York. Vertical scale is exaggerated.

Natural Gas Resource Potential

The USGS indicates that the Marcellus Shale may have a mean undiscovered natural gas resource potential of nearly 2 tcf. Considering the total extent of Devonian/Ohio basin shales (which include the Marcellus formation), an estimated 12 tcf may be present, although not all of it may be economically recoverable. (See **Figure 7**.) The USGS estimate, however, is considerably lower than a 2008 "conservative" estimate of 516 tcf made by geoscience professors Terry Engelder (Pennsylvania State University) and Gary Lash (State University of New York). The two professors believe that there could be at least 50 tcf of gas technically recoverable from the Marcellus Shale.[30]

Typically, thicker shales with greater organic material yield more gas, and thus, are more economically desirable to produce. Shale in northeast Pennsylvania and southeast New York has these characteristics and produces dry natural gas. Shale in western Pennsylvania and New York produces a wetter gas that contains petroleum liquids.

StatoilHydro assumes that each well drilled in the Marcellus may have an average estimated ultimate recovery (EUR) of 3.1 bcf.[31] The EUR per well will depend on the length of the horizontal well drilled, the number of fractures in the well, and the quality of the shale. StatoilHydro assumes that each horizontal well would measure 3,000 feet in length with 6 hydraulic fractures intervals in each well. An average well might produce for upwards of 60 years.

---

[30] "Marcellus Shale Play's Vast Potential Creating Stir in Appalachia," by Terry Engelder and Gary Lash, Professors of Geosciences, The American Oil and Gas Reporter, May 2008.

[31] StatoilHydro, "StatoilHydro's acquisition of 32.5% of Chesapeake interest in the Marcellus shale," press release, 2008, http://www.statoilhydro.com/en/NewsAndMedia/News/2008/Downloads/Frequently%20asked%20questions.pdf.

Figure 7. Devonian Shale Undiscovered Resource Potential

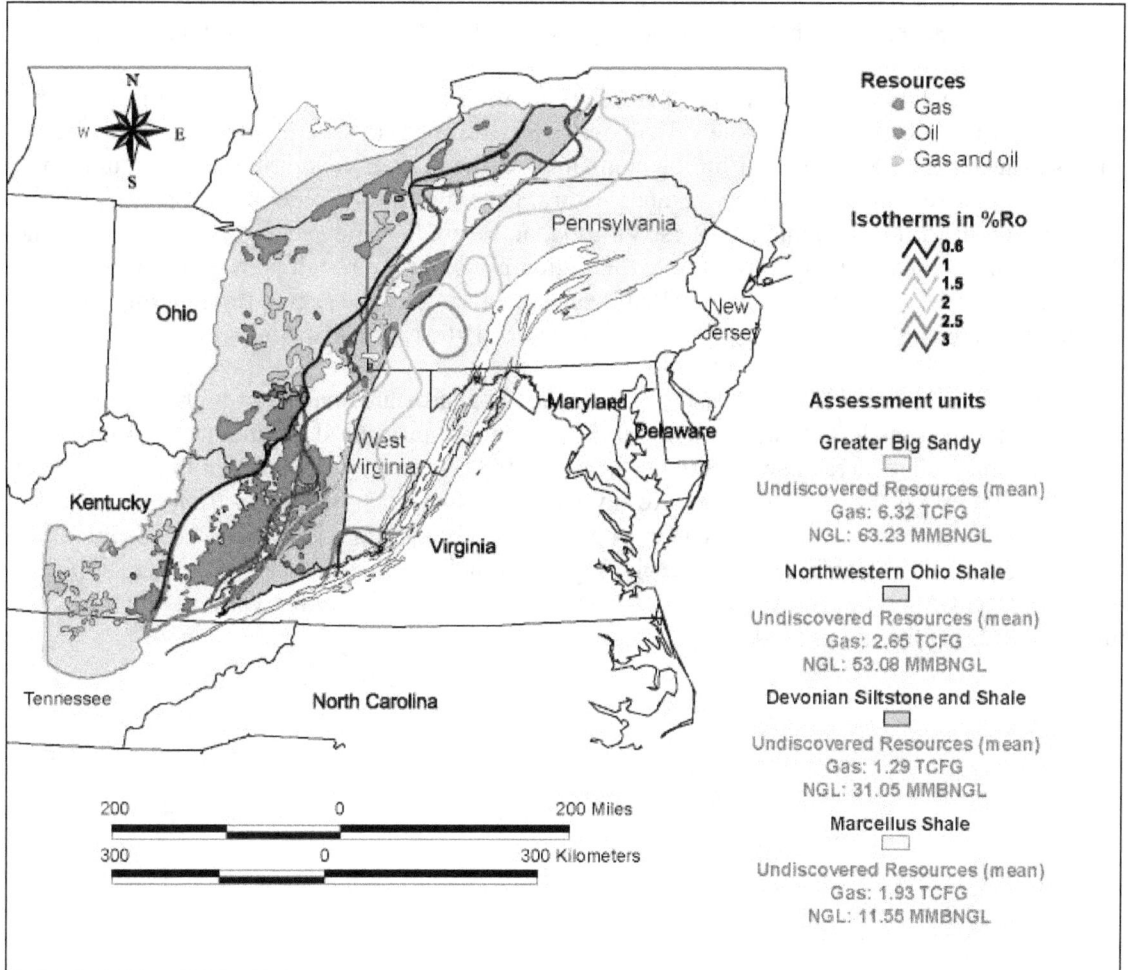

Source: USGS Open File Report 2005-1268.

Notes: NGL: natural gas liquids include butane and propane.

Northeast Regional Natural Gas Supply and Demand

In 2007, the northeast region consumed roughly 4 tcf of natural gas.[32] New York led the region in consumption with over 1.19 tcf. The United States as a whole consumed more than 23 tcf.[33] The region produced roughly 580 bcf of natural gas from some 113,000 operating gas wells. Pennsylvania and West Virginia combined made up nearly 89% of the production, with New York and Virginia making up the balance. In summary, the region consumes about seven times as much natural gas as it currently produces. The 50-tcf of gas, estimated by Engelder and Lash as

---

[32] U.S. DOE Energy Information Administration, *Natural Gas Consumption by End Use*, http://tonto.eia.doe.gov/dnav/ng/ng_cons_sum_dcu_nus_a.htm.

[33] Reported by EIA as 23,047,641 million cubic feet, http://tonto.eia.doe.gov/dnav/ng/ng_cons_sum_dcu_nus_a.htm.

---

technically recoverable from the Marcellus Shale, would be sufficient to supply the region for 13 years at the current rate of consumption. Taking the entire 513-tcf resource-potential into account and assuming that recovery methods would improve might extend the supply further.

Currently, 20 interstate natural gas transmission pipelines serve the northeast region of the United States (see **Figure 8**).[34] This pipeline system delivers natural gas to several intrastate natural gas pipelines and at least 50 local distribution companies in the region. In addition to the natural gas produced in the region, several long-distance natural gas transmission pipelines supply the region from the southeast into Virginia and West Virginia, and from the Midwest into West Virginia and Pennsylvania. Canadian imports come into the region principally through New York, Maine, and New Hampshire. Liquefied natural gas (LNG) supplies also enter the region through import terminals located in Massachusetts, Maryland, and New Brunswick, Canada.

The natural gas produced from the eastern portion of the Marcellus Shale is of high enough quality that it requires little or no treatment for injection into transmission pipelines.[35] A gas transmission pipeline already serves the northeast United States. The Millennium Pipeline project in southern New York could accommodate any increased shale gas production from New York and parts of Pennsylvania to serve the natural gas needs of the region. However, the terrain in West Virginia presents an obstacle to developing additional pipeline capacity and other support infrastructure. Gas producers would also have to construct an extensive network of gathering pipelines to bring the gas out of the well fields.

---

[34] U.S. DOE Energy Information Administration, *Natural Gas Pipelines in the Northeast Region*, http://www.eia.doe.gov/pub/oil_gas/natural_gas/analysis_publications/ngpipeline/northeast.html.

[35] The natural gas received and transported by the major intrastate and interstate mainline transmission systems must meet the quality standards specified by pipeline companies in the "General Terms and Conditions (GTC)" section of their tariffs. These quality standards vary from pipeline to pipeline and are usually a function of a pipeline system's design, its downstream interconnecting pipelines, and its customer base. In general, these standards specify that the natural gas: be within a specific Btu content range (1,035 Btu per cubic foot, +/- 50 Btu); be delivered at a specified hydrocarbon dew point temperature level (below which any vaporized gas liquid in the mix will tend to condense at pipeline pressure); contain no more than trace amounts of elements such as hydrogen sulfide, carbon dioxide, nitrogen, water vapor, and oxygen; and be free of particulate solids and liquid water that could be detrimental to the pipeline or its ancillary operating equipment. U.S. DOE Energy Information Administration, *Natural Gas Processing: The Crucial Link Between Natural Gas Production and Its Transportation to Market,* http://www.eia.doe.gov/pub/oil_gas/natural_gas/feature_articles/2006/ngprocess/ngprocess.pdf.

Figure 8. Northeast Region Natural Gas Pipeline Network

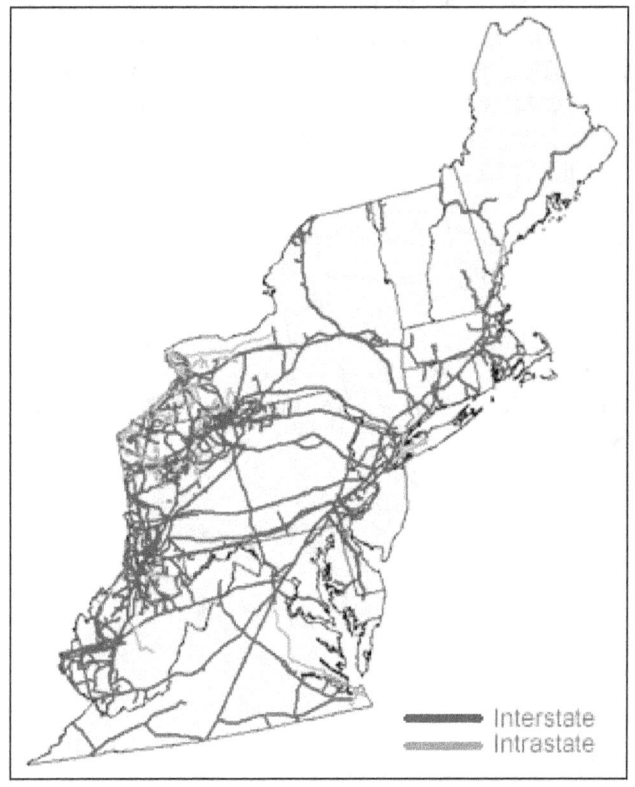

Source: Energy Information Administration, Office of Oil & Gas, Natural Gas Division, Gas Transportation Information System, http://www.eia.doe.gov/pub/oil_gas/natural_gas/analysis_publications/ngpipeline/northeast.html.

Notes: Includes Connecticut, Delaware, Massachusetts, Maine, New Hampshire, New Jersey, New York, Pennsylvania, Rhode Island, Virginia, and West Virginia.

Groundwater Resource Issues

The U.S. Geological Survey identifies three principal hydrogeological environments overlying the Marcellus Shale: (1) glacial sand and gravel aquifers in New York, northern Pennsylvania, and northeastern Ohio, (2) valley-and-ridge carbonate rock and other aquifers in Pennsylvania and eastern West Virginia, and (3) Mississippian aquifers in northern Pennsylvania and northeastern Ohio. These aquifer systems are important supplies of fresh water for communities, especially in more rural areas, although in general most residents of these states obtain their drinking water from surface water sources. Typically, these aquifers are much closer to the ground surface than the Marcellus Shale, which can be thousands of feet deep. The groundwater wells in these states may reach only several hundred feet in depth.

The layers of rocks separating most fresh water aquifers from the Marcellus Shale are typically siltstones and shales layered with minor sandstones and limestone. Siltstones and shales generally act as barriers to fluid flow. These intervening layers of rocks can be several thousand feet thick in the eastern and northern portions of the area where the Marcellus Shale is deepest. On the western and southern portions of the area, the Marcellus Shale is shallower, as separated by a thinner package of siltstones and shales.

Many of the surficial sand and gravel aquifers form valley-fill deposits, in low-lying areas or stream valleys, and are recharged by precipitation that runs off surrounding, less permeable uplands.[36] As such, they would be particularly susceptible to leaky surface impoundments or careless surface disposal because of the relatively short distance and travel time from the land surface to the top of the water table. New York, for example, has deemed these unconsolidated sand and gravel aquifers "primary" or "principal" aquifers, which are highly productive and presently used as a significant source of water, or are a potentially abundant water supply.[37]

The EPA defines an aquifer that supplies at least 50 % of the drinking water consumed in the area overlying the aquifer as a "sole or principal source aquifer" (referred to, for convenience, as sole source aquifers).[38] Those who depend on the aquifer for drinking water may have no alternative drinking water source (physically, legally or economically). The EPA has designated at least four sole source aquifers in New York (EPA Region II) and one in Pennsylvania (Region III) that overlie the Marcellus Shale.[39] (See **Figure 9**). It is uncertain whether these areas would be targets for Marcellus Shale development.

---

[36] Henry Trapp, Jr. and Marilee A. Horn, "U.S. Geological Survey Ground Water Atlas of the United States," HA 730-L (1997).

[37] See New York State, Department of Environmental Conservation, "Primary & Principal Aquifers," at http://www.dec.ny.gov/lands/36119.html.

[38] U.S. Environmental Protection Agency, *Sole Source Aquifer Protection Program*, http://cfpub.epa.gov/safewater/sourcewater/sourcewater.cfm?action=SSA.

[39] Cattaraugus Creek Basin Aquifer, Cortland Homer-Preble Aquifer System, Clinton Street-Ballpark Valley Aquifer System, and Schenectady/Niskanyuna Aquifers in New York, and Steven Valleys Aquifer in Pennsylvania.

Figure 9. Sole Source Aquifers in the Appalachian—Northeast Region

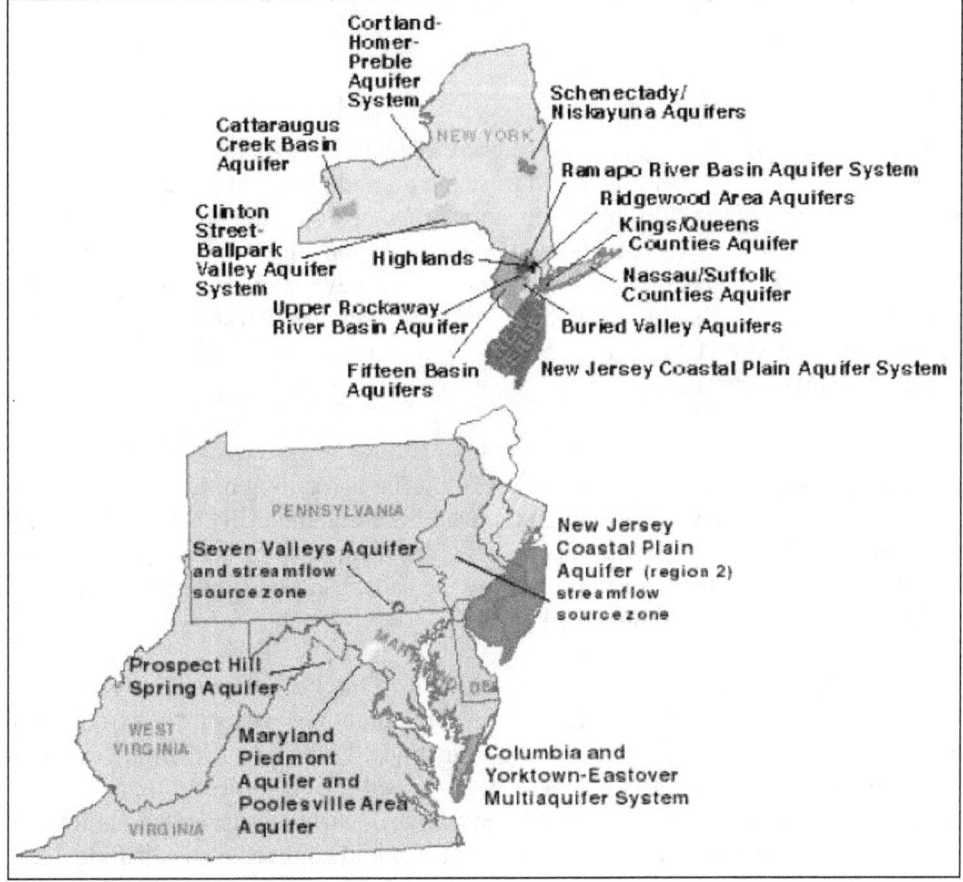

Source: U.S. EPA, *Sole Source Aquifer Protection Program*, Regions II and III, http://cfpub.epa.gov/safewater/sourcewater/sourcewater.cfm?action=SSA (modified by CRS).

Notes: EPA defines a sole or principal source aquifer as an aquifer that supplies at least 50 % of the drinking water consumed in the area overlying the aquifer.

Drilling and Development Technology

In their early stages of development, conventional petroleum reservoirs depend on the pressure of their gas-cap and oil-dissolved gas to lift the oil to the surface (i.e., gas drive). Water trapping the petroleum from below also exerts an upward hydraulic pressure (water-drive). Petroleum reservoirs produced by the pressure of their natural gas and water drives are thus termed "conventional drive." As a reservoir's production declines, lifting further petroleum to the surface requires pumping—giving rise to the term "artificial lift." In the late 1940s, drilling companies began inducing hydraulic pressure in wells to fracture the producing formation. This stimulated further production by effectively increasing a well's contact with a formation. Advances in directional drilling technology now enable wells to deviate from nearly vertical to extend horizontally into the reservoir formation, which increases the well's contact with the reservoir. Directional drilling technology also enables drilling a number of wells from a single well pad, thus cutting costs while reducing environmental disturbance. Combining hydraulic fracturing with directional drilling has opened up production of tighter (less permeable) petroleum and natural gas reservoirs, and in particular, unconventional gas shales like the Marcellus.

Drilling

Well drilling has progressed from an art to a science. Originally, drillers used "cable-tool" rigs and a percussion bit. The drill operator would raise the bit and release it to pulverize the sediment. From time to time, the driller would stop and "muck out" the pulverized rock cuttings to advance the well. Though time-consuming, this method was simple and required minimal labor. Some drillers still use this method for water-wells and even some shallow gas-wells. The introduction of rotary-drill rigs at the beginning of the 20th century marked a big advance in drilling, particularly with the development of the "tri-cone rotary bit."[40] This method, as the name implies, uses a weighted rotating bit to penetrate the sediment (see **Figure 10**).

The key to a rotary drill's speed is the relative ease of adding new sections of drill pipe (or drill string) while the drill-bit continues turning. Circulating fluids (drilling mud) down through the center of the hollow drill pipe and up through the well bore lifts the drill cuttings to the surface. Modern drill bits studded with industrial diamonds gives them an abrasive property to grind through any rock type. However, from time to time the drill string must be removed (a process termed "tripping") to replace the dulled drill bit.

To function properly, drilling fluids must lubricate the drill bit, keep the well bore from collapsing, and remove cuttings. The weight of the mud column prevents a "blow-out" from occurring when encountering high-pressure reservoir fluids. Drillers base the mud's composition on natural bentonite clay, a "thixotropic" material that is solid when still and fluid when disturbed. This essential rheological property keeps the drill cuttings suspended in the mud. The mud's chemistry and density must be carefully monitored and adjusted as the drilling deepens (for example, adding a barium compound increases mud density). "Mud pits," excavated adjacent to the drill rig provide a reservoir for mixing and holding the mud. The mud pits also serve as settling ponds for the cuttings. At the completion of drilling, the mud may be recycled at another drilling operation, but the cuttings will be disposed of in the pit. Several environmental concerns over drilling stem from the (hazardous) composition of the drilling mud and cuttings, and the potential for mud pits overflowing and contaminating surface water.

---

[40] Howard Hughes, Jr. of the Hughes Tool Company developed the modern tri-cone rotary bit. His father, Howard Robert Hughes, Sr. had invented the bit's ancestor, a two-cone rotary bit.

Figure 10. Rotary Drilling Rig

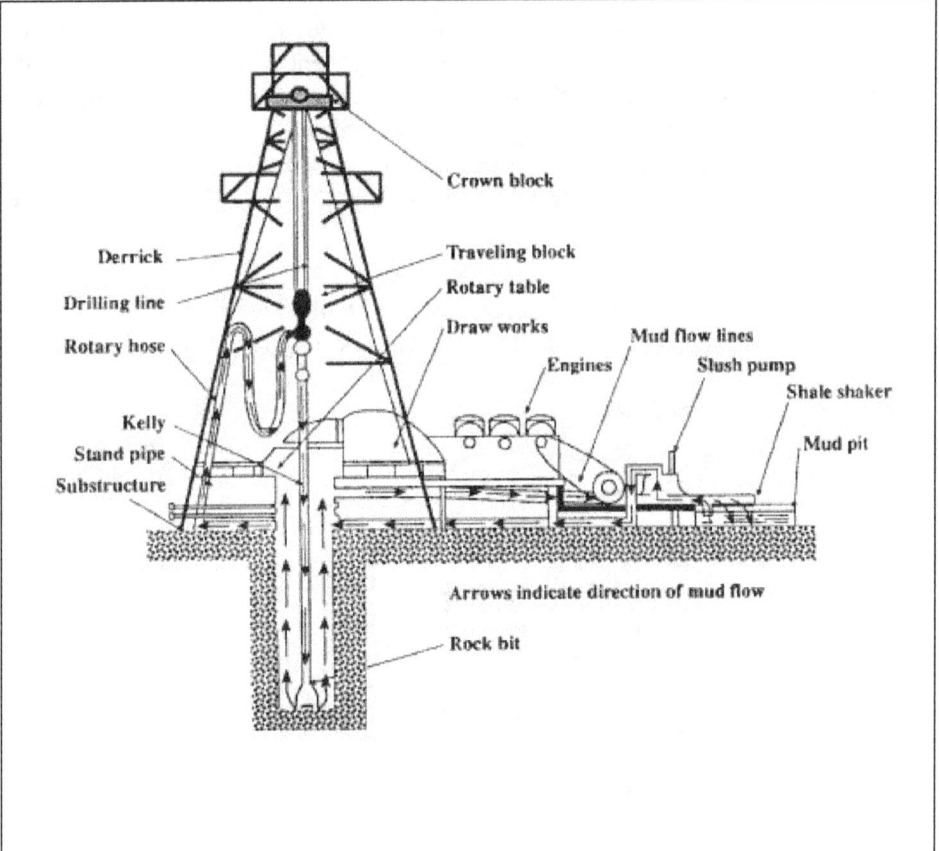

Source: Bureau of Land Management, http://www.blm.gov/rmp/wy/rawlins/documents/RMO/figures/FigureA-5.gif.

The most recent advance in drilling is the ability to direct the drill bit beyond the region immediately beneath the drill rig. Early directional drilling involved placing a steel wedge down-hole (whipstock) that deflected the drill toward the desired target, but lacked control and consumed time. Advances such as steerable down-hole drill motors that operated on the hydraulic pressure of the circulating drilling mud offered improved directional control. However to change drilling direction the operator had to halt drill string rotation in such a position that a bend in the motor pointed in the direction of the new trajectory (referred to as the sliding mode). Newer rotary steerable systems introduced in the 1990s eliminated the need to slide a steerable down-hole motor.[41] The newer tools drill directionally while continuously rotated from the surface by the drilling rig. This enables a much more complex, and thus accurate, drilling trajectory. Continuous rotation also leads to higher rates of penetration and fewer incidents of the drill-string sticking. (See **Figure 11**.)

---

[41] Schlumberger, *Better Turns for Rotary Steerable Drilling: Overview*, http://www.slb.com/content/services/resources/oilfieldreview/ori002/01.asp?.

Directional drilling offers another significant advantage in developing gas shales. In the case of thin or inclined shale formations, a long horizontal well increases the length of the well bore in the gas-bearing formation and therefore increases the surface area for gas to flow into the well. However, the increased well surface (length) is often insufficient without some means of artificially stimulating flow. In some sandstone and carbonate formations, injecting dilute acid dissolves the natural cement that binds sand grains thus increasing permeability. In tight formations like shale, inducing fractures can increase flow by orders of magnitude. However, before stimulation or for that matter production can take place, the well must be completed and cased.

Figure 11. Directional Drilling

Steerable Down-hole Motor vs. Rotary Steerable System

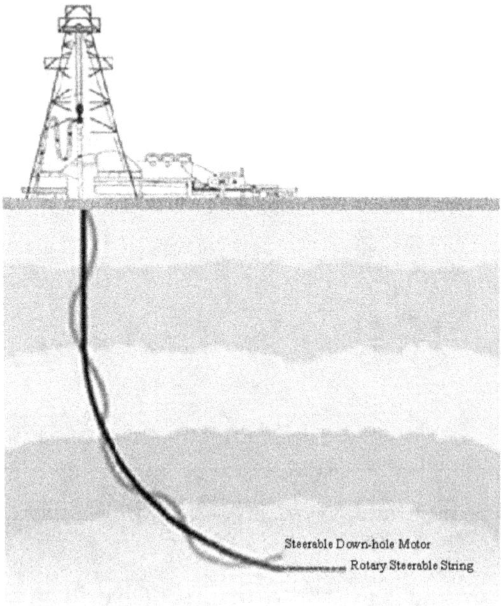

Steerable Down-hole Motor
Rotary Steerable String

Source: Schlumberger, modified by CRS.

Well Construction and Casing

Commercial gas and oil, and municipal water-supply wells have in common a series of telescoping steel well casings that prevent well-bore collapse and water infiltration while drilling. The casing also conducts the produced reservoir fluids to the surface (see **Figure 12**). A properly designed and cemented casing also prevents reservoir fluids (gas or oil) from infiltrating the overlying groundwater aquifers.

During the first phase of drilling, termed "spuding-in," shallow casing installed underneath the drilling platform serves to reinforce the ground surface. Drilling continues to the bottom of the water table (or the potable aquifer), at which point the drill string is "tripped" out (removed) in order to lower a second casing string, which is cemented-in and plugged at the bottom. Drillers use special oil-well cement that expands when it sets to fill the void between the casing and the wellbore.

Surface casing and casing to the bottom of the water table prevents water from flooding the well while also protecting the groundwater from contamination by drilling fluids and reservoir fluids. (The initial drilling stages may use compressed air in place of drilling fluids to avoid contaminating the potable aquifer.) Drilling and casing then continue to the "pay zone"—the formation that produces gas or oil. The number and length of the casings, however, depends on the depth and the properties of the geologic strata.

Figure 12. Hypothetical Well Casing

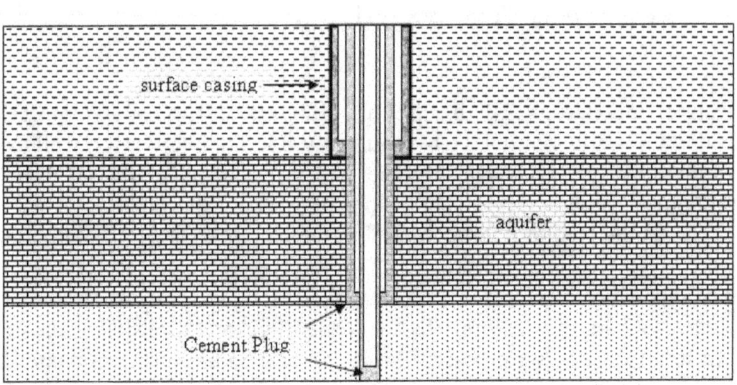

Source: CRS.

Notes: No Scale.

After completing the well to the target depth and cementing-in the final casing, the drilling operator may hire an oil-well service company to run a "cement evaluation log." An electric probe lowered into the well, measures the cement thickness. The cement evaluation log provides the critical confirmation that the cement will function as designed—preventing well fluids from bypassing outside the casing and infiltrating overlying formations.

Absent any cement voids, the well is ready for completion. A perforating tool that uses explosive shape charges punctures the casing sidewall at the pay zone. The well may then start producing under its natural reservoir pressure or, as in the case of gas shales, may need stimulation treatment.

Both domestic-use gas wells and water wells are common throughout regions experiencing recent shale gas development. In the absence of regulation, domestic-use wells (gas or water) may not meet standard practices of construction. If the wellhead of a water supply well is improperly sealed, for example, surface water may infiltrate down along the casing exterior and contaminate the drinking-water aquifer. Some domestic water wells have also produced natural gas, and some shallow gas wells have leaked into nearby building foundations. To avoid some of these problems, Pennsylvania is instituting regulations that require a minimum 2,000-foot setback between a new gas-well and an existing water-well.

Hydraulic Fracturing

Despite some shales' abundant natural gas content, they do not produce gas freely. Economic production depends on some means of artificially stimulating shale to liberate gas. In the late 1940s, Texas oil fields responded to fluids pumped down wells under pressures high enough to fracture stimulated the producing formation. Hydraulic fracture stimulation treatments have been adapted to tight gas formations such as the Barnett Shale in Texas, and more recently the Marcellus Shale.

Typical "frac" treatments or frac jobs (as commonly referred to) are relatively large operations compared to some drilling operations. The oilfield service company contracted for the work may take a week to stage the job, and a convoy of trucks to deliver the equipment and materials needed (see **Figure 13**).

A company involved in developing Texas gas shale offered the following description of a frac job:[42]

> Shale gas wells are not hard to drill, but they are difficult to complete. In almost every case, the rock around the wellbore must be hydraulically fractured before the well can produce significant amounts of gas. Fracturing involves isolating sections of the well in the producing zone, then pumping fluids and proppant (grains of sand or other material used to hold the cracks open) down the wellbore through perforations in the casing and out into the shale.

> The pumped fluid, under pressures up to 8,000 psi, is enough to crack shale as much as 3,000 ft in each direction from the wellbore. In the deeper high-pressure shales, operators pump slickwater (a low-viscosity water-based fluid) and proppant. Nitrogen-foamed fracturing fluids are commonly pumped on shallower shales and shales with low reservoir pressures.

---

[42] Schlumberger, Inc., *Shale Gas: When Your Gas Reservoir is Unconventional, So is Our Solution.* http://www.slb.com/media/services/solutions/reservoir/shale_gas.pdf.

Figure 13. Hydraulic Fracture Job
Marcellus Shale Well, West Virginia

Source: Chesapeake Energy Corporation, 2008.

Notes: The yellow frac tanks in the foreground and along the tree line hold water, the red tanker holds proppant; hydraulic pumps are in the center.

Fracturing Fluids

Fracturing fluid functions in two ways: opening the fracture and transporting the "propping" agent (or proppant) the length of the fracture.[43] As the term propping implies, the agent functions to prop or hold the fracture open. The fluid must have the proper viscosity and low friction pressure when pumped, it must breakdown and cleanup rapidly when treatment is over, and it must provide good fluid-loss control (not dissipate). The fluid chemistry may be water-based, oil-based or acid-based depending on the properties of the formation. Water-based fluids (sometimes referred to as slickwater) are the most widely used (especially in shale formations) because of their low cost, high performance, and ease of handling. Some fluids may also include nitrogen and carbon dioxide to help foaming. Oil-based fluids find use in hydrocarbon bearing formations susceptible to water damage, but are they expensive and difficult to use. Acid-based fluids use hydrochloric acid to dissolve the mineral matrix of carbonate formations (limestone and dolomite) and thus improve porosity; the reaction produces inert calcium chloride salt and carbon dioxide gas.

---

[43] "Chapter 7 - Fracturing Fluid Chemistry and Proppants," in *Reservoir Stimulation*, ed. Michael J. Economides and Kenneth G. Nolte, 3rd ed. (John Wiley & Sons, LTD, 2000).

Water-based fluids consist of 99% water with the remainder made up of additives. The initial fracturing stage may use hydrochloric acid (HCl) to cleanup the wellbore damage done during drilling and cementing. Gelling agents, based on water-soluble polymers such as vegetable-derived guar gum, adjust frac fluid viscosity. The most widely used additives for breaking down fluid viscosity after fracturing are oxidizers such as ammonium (NH^{+4}), potassium, and sodium salt of peroxydisulfate ($S2O_8^{-2}$); enzyme breakers may be based on hemicellulase (actually a mixture of enzymes which can hydrolyze the indigestible components of plant fibers). Silica flour serves as good fluid-loss additive. Biocides added to polymer-containing fluids prevent bacterial degradation (as the polysaccharides (sugar polymer) used to thicken water are an excellent food source for bacteria). Methanol (an alcohol) and sodium thiosulfate ($Na_2S_2O_3$) are commonly used stabilizers added to prevent polysaccharide gels degrading above temperatures of 200°F.

It is important to note that the service companies adjust the proportion of frac fluid additives to the unique conditions of each well. The Occupational Safety and Hazard Administration (OSHA) requires that material safety data sheets (MSDS) accompany each chemical used on the drill site, but the proportion of each chemical additive may be kept proprietary.[44]

Proppants hold the fracture walls apart to create conductive paths for the natural gas to reach the wellbore. Silica sands are the most commonly used proppants. Resin coating the sand grains improves their strength.

Hydraulic Fracture Process

Fracture treatments are carefully controlled and monitored operations that proceed in stages. Before beginning a treatment, the service company will perform a series of tests on the well to determine if it is competent to hold up to the hydraulic pressures generated by the fracture pumps.

In the initial stage, an HCl solution pumped down the well cleans up residue left from cementing the well casing. Each successive stage pumps discrete volumes of fluid (slickwater) and proppant down the well to open and propagate the fracture further into the formation. The treatment may last upwards of an hour or more, with the final stage designed to flush the well. Some wells may receive several or more treatments to produce multiple fractures at different depths, or further out into the formation in the case of horizontal wells.

A single fracture treatment may consume more than 500,000 gallons of water.[45] Wells subject to multiple treatments consume several million gallons of water. An Olympic-size swimming pool (164 ft x 82 ft x 6 ft deep) holds over 660,000 gallons of water, for comparison, and the average daily per capita consumption of fresh water (roughly 1,430 gallons per day) is 522,000 gallons over one year.[46]

---

[44] 29 C.F.R. §§ 1910 Subpart Z, Toxic and hazardous substances.

[45] Modern Shale Gas Development in the United States: A Primer, pp. 58-59.

[46] U.S. Geological Survey, *Summary of water use in the United States, 2000*, http://ga.water.usgs.gov/edu/wateruse2000.html.

The high injection pressure not only opens and propagates the fracture but also drives fluid into the shale's pore spaces. A high volume of fluid also remains in the fracture that will impede gas flow to the well if not pumped out. The subsequent "flowback" treatment performed, attempts to recover as much of the remaining fluid as possible without removing the proppants. The "flowback" water pumped out of the well may be high in dissolved salts and frac chemicals, however, making it unsuitable for beneficial use, and requiring treatment for disposal. After the well begins producing gas, it may also produce more flowback water. Flowback disposal presents environmental issues, as discussed in the "Surface Water Quality Protection" section below.

Fracture Geometry

The fracture is ideally represented by a vertical plane that intersects the well casing (**Figure 14**). It does not propagate in a random direction, but opens perpendicular to the direction of least stress underground (which is nearly horizontal in orientation).

Figure 14. Idealized Hydraulic Fracture

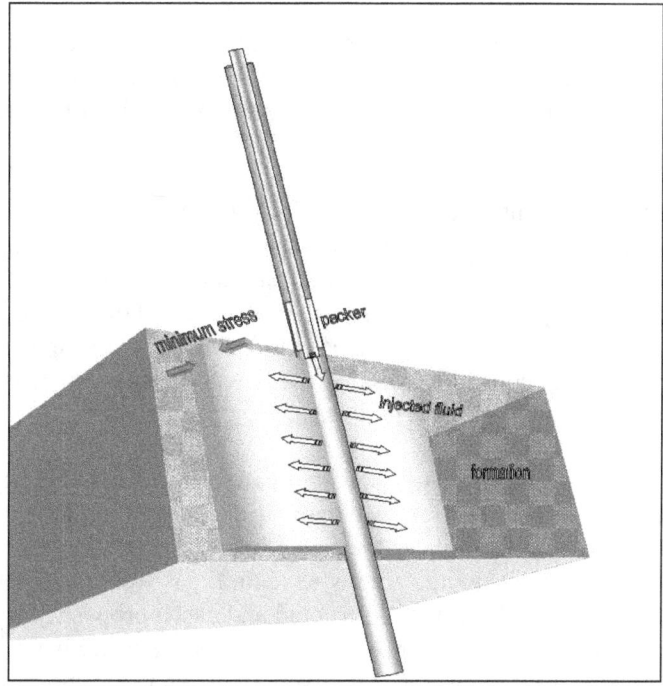

Source: CRS.

Notes: Fracture opens perpendicular to direction of minimum stress.

Hydraulic fracturing attempted at too shallow a depth develops an undesirable horizontal fracture. In the case of deep shales, such as the Marcellus, this appears unlikely. The fracture may extend outward several thousands of feet, but the fracture height is limited to the thickness of the shale formation (and controlled by the perforated zone of the well).

Fracturing Risks to Groundwater

The geologic environment that led to the deposition of the Marcellus Shale, and the overlying layers of siltstone, shale, sandstone, and limestone have kept gas from the Marcellus Shale confined at depth, and prevented it from naturally migrating upwards toward the surface. The process of developing a shale gas well (drilling through an overlying aquifer, stimulating the well via hydraulic fracturing, completing the well and producing the gas) is an issue of concern for increasing the risk of groundwater contamination. Typically, well drilling and completion practices, as described above, require sealing the well by casing throughout the aquifer interval. A properly designed and cased well will prevent drilling fluids, hydraulic fracturing fluids, or natural gas from leaking into the permeable aquifer and contaminating groundwater. The casing also prevents groundwater from leaking into the well where it could interfere with the gas production process.

An environmental concern raised by hydraulic fracturing is the possibility of introducing contaminants into aquifers. Hydraulic fracturing does induce new fractures into shale, and can propagate fractures thousands of feet along the bedding plane of a shale formation. The potential for propagating fractures to an overlying aquifer may depend on the depth separating the two. However, engineers designing and conducting frac jobs have a strong incentive to limit the fractures to the height of the gas-producing shale zones. Any fracture propagated to an overlying aquifer could allow water to flow down into the gas-producing portion of the shale, which could significantly hamper gas production.

Multiple frac treatments may pump as much as 2 to 3 million gallons of fluid down a well. Afterwards the well operator recovers a large proportion of these fluids by pumping them out of the well, and disposes of them through waster-water treatment plants or by other means as discussed below. However, any drilling fluids or frac fluids spilled on the ground surface could infiltrate downwards to shallow groundwater. This could pose a risk to surficial aquifers composed of very permeable unconsolidated sand and gravel deposits, such as those in northern Pennsylvania and southern New York (see **Figure 9**).

Another potential issue could be groundwater contamination from poorly constructed water wells. Generally, drinking water wells are shallower than natural gas wells, and their casing may not extend their entire depth. This is particularly the case for domestic water wells that may not be subject to the same level of oversight and scrutiny as natural gas well construction. A water well that is not cased from the surface, or is not constructed and cased properly, might allow contaminated water to flow from the ground surface and enter the water well, possibly compromising the quality of drinking water in the well and even the drinking water aquifer itself. In such instances, and particularly where natural gas drilling and stimulation activities are nearby, leaky surface impoundments or careless surface disposal of drilling fluids at the natural gas operation could increase the risk of contaminating the nearby water well.

Contaminated surface water unrelated to drilling could also contaminate improperly cased or constructed water well. For example, a leaky septic system, or improper disposal of domestic refuse such as car batteries or used oil, could leak from the surface into the water well. If this is the case, a dispute could ensue as to who is responsible for contaminating the water well. Resolving the dispute could involve a hydrogeological investigation to prove or disprove any linkage between natural gas development activities and water well contamination, often at considerable expense and with an uncertain outcome, given the complexity of groundwater flow at most sites.

Leasing Issues for Gas Development

The Marcellus Shale gas formation has generated considerable interest within the past two years on state-owned and private lands throughout the Appalachian region in West Virginia, Pennsylvania, and New York. Landowners who also own the subsurface mineral rights may lease their land for the development of those minerals or sell the surface or mineral rights to qualified buyers for development.[47] If the landowner decides to lease the land for mineral development, the owner may negotiate a lease agreement or contract. Common features of oil and gas leases today include signing bonuses (or up front "bonus bid" payments for competitive lease sales), royalties, rents, primary lease terms and conditions for lease renewal.

In the case of mineral rights severed from surface rights, a split-estate is established. Under a split-estate, the mineral owner has the right to reasonable access to recover the mineral. The surface owner typically has the right to protection from "unreasonable encroachment and damage" and can usually negotiate compensation for the use of the surface and for damages. On split-estates that involve private surface owners and federally owned minerals, a federal lessee must meet one of the following conditions: a surface agreement; written consent or a waiver from the private surface owner for access to the leased lands; payment for loss or damages; or the execution of a bond not less than $1,000.[48] Split-estate scenarios are common but a comprehensive discussion on the subject is beyond the scope of this report.

The state-by-state sections below describe how compensation to landowners atop the Marcellus Shale has evolved over the past few years. The discussion focuses primarily on bonus payments and royalties received by state and private landowners today compared to what they received just a few years ago. Rents are not part of the discussion because nearly all of the information available reports on signing bonuses and royalties. Further, lease agreements often roll signing bonuses into rents, and call for rents paid upfront or paid quarterly as a "delay rental." Rents appear to be much less significant to small landowners who lease a few acres. On state and private lands, as with federal lands, lessees usually pay rent until production commences and then switch to paying royalties on the value of production.

New York

In 1999, signing bonuses for leases on private lands were around $5 per acre, with royalties of 12½%. During that same time, the state received between $15 and $600 per acre in bonus payments. In today's climate, bonus payments are as much as $3,000 per acre on privately held lands with royalty rates between 15% and 20%. Lease terms are typically three to five years with renewal clauses for continued production and shut-in wells. Fewer than 10 years ago, there were only about five or six companies involved in securing leases to develop gas shale projects. By 2008 their number rose to about four dozen companies (including Exxon/Mobil, BP, and Conoco Phillips).

---

[47] The term "mineral" refers to economically recoverable resources including natural gas.

[48] This information was provided in a BLM Instructional Memorandum, No. 2003-131. Also see 43 CFR 3104, 43 CFR 3814 and 48 FR 48916 (1983) for further details.

Pennsylvania

In 2002, companies were paying signing bonuses of $25 to $50 per acre and agreeing to $2 per acre annual rentals and 12½ % royalty on 10-year leases on state-owned land. Interest has grown dramatically since that time, and firms are now paying as much as $2,500 per acre in bonus payments, 16% royalty and annual rental rates of $20 per acre for years 2 through 5 and $35 per acre in years 6 through 10. Private landowners have typically issued shorter-term leases of 5 to 7 years. Private owners received about $12 per acre in signing bonuses and a 12½% royalty rate in 2003. In 2008, lessees are now paying private landowners signing bonuses of nearly $2,900 per acre with 17% to 18% royalty rates on 5- through 7-year leases. Many private leases often calculate rents as part of the signing bonus and sometimes pay them upfront or over time.

West Virginia

Some landowners in West Virginia have seen their bonus bids climb from $5 per acre in 2007 and early 2008, to more recent bonus payments of $1,000 to $3,000 per acre. Royalty rates have increased from 12½% through 16% to 18%. Rents are often included in the signing bonus or sometimes paid out in the form of a "delay rental."

State Summary

In general, Marcellus gas lease rates range for state and private lands from a low of 12½% in West Virginia to high of 20% in New York (**Table 1** and **Table 2**), but these rates are less than those paid in Texas for similar gas shale tracts. All three of Marcellus Shale states reported above have shown significant increases in the amounts paid as signing bonuses and increases in royalty rates. There are lease sales that the Natural Gas Leasing Tracking Service reported with signing bonuses as low as $100 to $200 per acre, owing to greater uncertainty and less interest among natural gas companies and/or the lack of information among landowners on what the land is worth.[49]

There are several landowner organizations that have formed in recent years to pool their land for leasing, advise landowners, and serve as information centers. Some groups seek out competitive bids from energy companies.[50]

---

[49] The Natural Gas Leasing Offer Tracking can be found at http://www.pagaslease.com/lease_tracking_2.php.

[50] Natural Gas /Oil Landowner Groups Directory, http://www.pagaslease.com/directory_public.php.

Table 1. Shale Gas Bonus Bids, Rents, and Royalty Rates on Selected State Lands

	Statutory Minimum or Standard Royal Rate	Royalty Rate Range	Bonus Bids (per acre)	Comments
West Virginia [a]	12½%	n.a.	n.a.	No state shale gas leases.
Pennsylvania [b]	12½%	12½% - 16%	$2,500	In many cases bonus bids were in the $25-$50 per acre range as recent as 2002. A royalty rate of 12.5% was most common.
New York [c]	12½%	15% - 20%	about $600	Bonus bids ranged from $15-$600 per acre around 1999-2000 and most royalty rates were at 12.5%.
Texas	12½%	25%	$350 - $400 (Delaware Basin) $12,000 (river tracts)	Bonus bids have been relatively consistent in recent times (within the past 5 years). Royalty rates were more common at 20%-25% about 5 years ago. Most state-owned lands are not considered to be among the best sites for shale gas development.

Source: CRS.

a. Personal communication with Joe Scarberry in the WV Department of Natural Resources, October 2008.

b. Personal communication with Ted Borawski in the PA Bureau of Forestry, who provided information on shale gas leases on both state and private lands, October 2008.

c. Personal communication with Lindsey Wickham of the NY Farm Bureau and Bert Chetuway of Cornell University, discussed lease sales on state and private land, October 2008.

Table 2. Shale Gas Bonus Bids, Rents, and Royalty Rates on Private Land in Selected States

	Royalty Rates Range	Bonus Bids (per acre)	Comments
West Virginia [a]	12½% –18 %	$1,000 – $3,000	Bonus payments were in the $5 per acre range as recently as 1-2 years ago. Royalty rates were 12½%
Pennsylvania	17% – 18%	$2,000 – $3,000	
New York	15% – 20%	$2,000 – $3,000	
Texas	25% – 28%	$10,000 – $20,000	Bonus bids were in the $1,000 range around 2000-2001. Royalty rates ranged from 20% to 25% range.

Source: CRS

a. Personal communication with David McMahon, Director of the WV Surface Owners Rights Organization, October 2008.

Lease Audit (Product Valuation and Verification)

Organizations involved with private-owned leases for shale gas development often recommend to lessors that they include provisions/clauses in their leases that require the producer to pay the wellhead price without deductions or to base the royalty on the gross proceeds at the well. Under most circumstances (in arms-length, third-party transactions), the wellhead is the point of first sale. Wellhead and spot prices are available from different sources including the EIA website. There are opportunities for a lessor to audit volume flows from the well, but more often, the producer will provide production and price data to the lessor. According to David McMahon of the West Virginia Surface Owners Rights Organization (WVSORO), there are major concerns among landowners that include the lessees deducting costs inappropriately, before the royalty is calculated.[51] This practice can deduct costs for transportation, compressors, and line loss from the wellhead price, thus reducing the royalty paid to the landowner. While metering production is typical, auditing production may be rare. The cost of an audit may be cost prohibitive for small landowners.

The state of Pennsylvania audits production from the top 100 wells of its natural gas leases using an independent auditor, or "meter truck companies" that work for the shale gas producers. The state of New York requires that natural gas producers meter production and make that information available upon request.[52] The West Virginia Department of Environmental Protection, Office of Oil and Gas, requires an annual production report from all oil and gas producers in the state.

Severance Taxes

State governments of Pennsylvania and New York have proposed a severance tax (included in the 2009-2010 budget) in order to generate revenue to address state concerns. Twenty-seven states producing natural gas (including West Virginia) already impose a severance tax. In Pennsylvania, based on the proposed severance tax, the Pennsylvania state government projected revenue increases of $107 million in 2009-2010, increasing to $632 million by 2013-2014.[53] A separate study on the economic impacts of Marcellus Shale gas development in Pennsylvania projected economic output of $3.8 billion, 48,000 jobs, and $400 million in state and local tax revenues in 2009.[54] The study also projected cumulative tax revenues reaching an estimated $12 billion (in present value dollars) in 2020. However, based on the proposed severance tax, state and local tax revenue estimates could be reduced by $880 million (in present value dollars) resulting from an estimated 30% reduction in natural gas drilling between 2009 and 2020 according to the study.

There are typically two types of severance taxes: an ad valorem or value-based tax or a unit of production-based tax. Individual states uniquely apply severance taxes to the extraction industries for the "privilege" of extracting natural resources (such as minerals and timber) in order to compensate residents for the irreplaceable loss from the extraction of the resource, sometimes described as a loss of future wealth. Revenue streams from severance taxes pay for the

---

[51] West Virginia Surface Owners' Rights Organization, http://www.wvsoro.org/.

[52] Energy Conservation Law, 23-0301-23-0305, Part 556.

[53] Pennsylvania Budget and Policy Center, *Responsible Growth, Protecting the Public Interest with a Natural Gas Severance Tax*, by Michael Wood and Sharon Ward, April 2009.

[54] Pennsylvania State University, College of Earth and Mineral Sciences, Department of Energy and Mineral Engineering, An *Emerging Giant: Prospects and Economic Impacts of Developing the Marcellus Shale Natural Gas Play*, by Timothy Considine, et al., August 5, 2009.

environmental and social costs associated with mining and extraction. The tax may also contribute to a state's general revenue fund.

The severance tax as other costs imposed on the oil and gas industry may affect long-term production rates. This variable cost, applied to the operator, only when production occurs, may slow extraction. A whole host of other factors, e.g., geology, price/demand, transportation costs, and capital formation, affect when and where natural gas is developed.

Federal Land Leasing and Restrictions to Leasing

Leasing federal public lands for oil and gas development is based on multiple-use/sustained yield Resource Management Plans (RMPs) prepared by the Bureau of Land Management (BLM) in the Department of the Interior. In accordance with those management plans, BLM offers tracts of public land with oil and gas potential for competitive leasing each quarter. The BLM administers oil and gas leasing and development on federally owned minerals both for BLM lands and on behalf of the U.S. Forest Service, which in turn administers the surface development within the National Forests and through its land planning process addresses surface occupancy concerns and approves or rejects any extraction of forest resources.

However, privately held mineral rights account for over 90% of the Allegheny National Forest in Pennsylvania and about 38% within the Monongahela National Forest in West Virginia. Finger Lakes National Forest in New York was withdrawn from the mineral leasing programs under section 370 the Energy Policy Act of 2005 (P.L. 109-58).

The Mineral Leasing Act and the Outer Continental Shelf Lands Act, as amended, requires that all public lands (onshore and offshore areas respectively) available for lease be offered initially to the highest responsible qualified bidder by competitive bidding. The objective of the competitive bid is to provide a "fair market value" return to the federal government for its resources. Under the two acts mentioned above, the federal government requires royalties from oil and gas producers (and other resource producers) on leasable federal lands and annual rents from non-producing lessees. Competitive oil and gas lease sales include bonus bids (upfront payments made to obtain a lease). The statutory minimum royalty rate for oil and gas leases on public lands is 12½% (also expressed as ⅛). Currently the royalty rate for onshore oil and gas leases is 12½%, while the rate for an offshore lease can range from 12½% to 18¾%. Annual rental rates established for onshore leases are at not less than $1.50 per acre for the first five years of a ten-year lease and not less than $2 per acre each year thereafter. The primary lease term for an onshore lease is 10 years and for an offshore lease is either 5, 8, or 10 years depending on the depth of the water. Annual rent paid on leases ends when production begins, then the lessor begins paying royalties on the value of production.

Of the approximately 5 million acres of federal land in the Appalachian basin (**Figure 15**), 46% (2.5 million acres) is not accessible for lease.[55] Based on resource estimates, these lands contain

---

[55] U.S Department of the Interior, Inventory of Onshore Federal Oil and Natural Gas Resources and Restrictions to Their Development, Phase III Inventory - Onshore United States, 2008, p. 225. http://www.blm.gov/pgdata/etc/medialib/blm/wo/MINERALS__REALTY__AND_RESOURCE_PROTECTION_/energy/EPCA_Text_PDF.Par.18155.File.dat/Executive%20Summary%20text.pdf

41% (984.7 bcf) of the federal natural gas in the basin. Another 42% (2.2 million acres) is accessible with restrictions on oil and gas operations beyond standard lease terms. These lands contain 45% (1.1 tcf) of the federal natural gas. The remaining 13% (691.7 thousand acres) that is accessible under standard lease terms contains 14% (346.7 bcf) of the federal natural gas. Most of the undiscovered gas resource (94%) will likely occur in continuous accumulations. Coal-bed natural gas accounts for about 13% of the total undiscovered continuous gas.

Figure 15. Federal Lands Overlying Natural Gas Resources of the Appalachian Basin

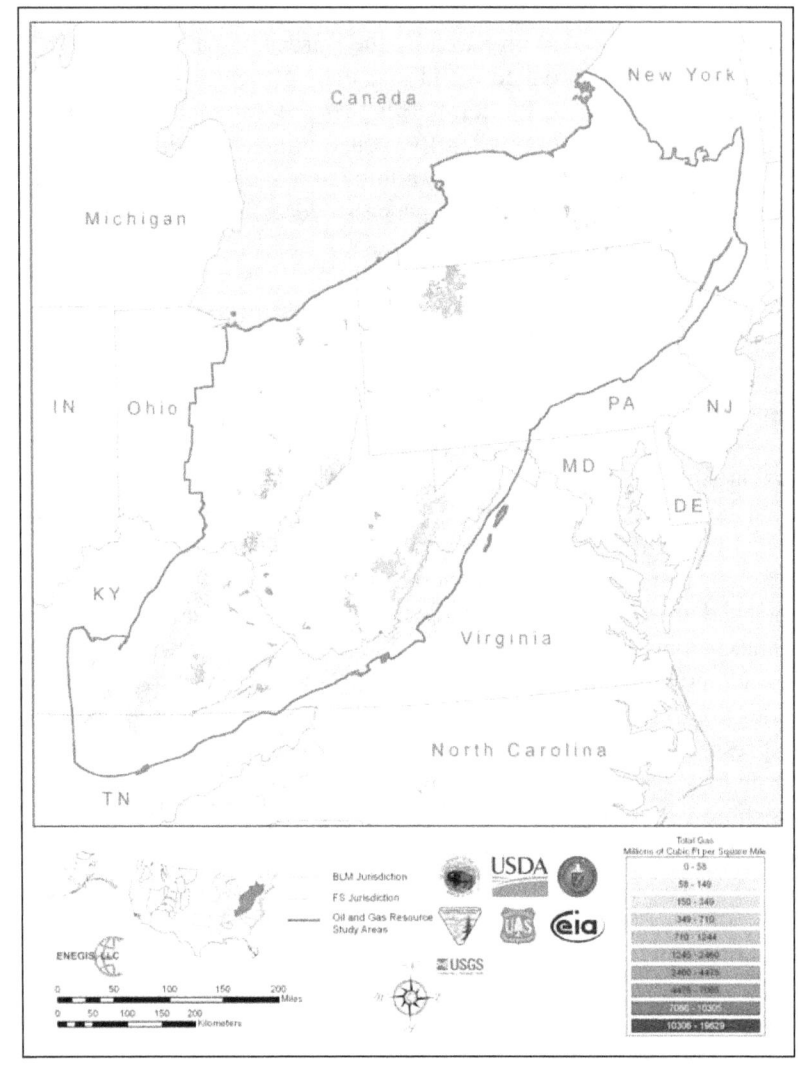

Source: U.S. Dept. of the Interior, *Inventory of Onshore Federal Oil and Natural Gas Resources and Restrictions to Their Development, Phase III Inventory,* 2008.

Notes: Approximately 5 million acres of federal lands in the Appalachian Basin region overlie natural gas resources.

Federal and State Laws and Regulations Affecting Gas Shale Development

Development of the Marcellus Shale will be subject to regulation under several federal and state laws. In particular, the large volumes of water needed to drill and hydraulically fracture the shale, and the disposal of this water and other wastewater associated with gas extraction may pose significant water quality and quantity challenges that trigger regulatory attention. As the U.S. Geological Survey noted in a recent publication, "Concerns about the availability of water supplies needed for gas production, and questions about wastewater disposal have been raised by water-resource agencies and citizens through the Marcellus Shale gas development region."[56] The following sections review key provisions of two relevant federal laws, the Safe Drinking Water Act (SDWA) and the Clean Water Act (CWA), and related state requirements.

Surface Water Quality Protection

As previously described, hydraulic fracturing involves injecting water, sand, and chemicals into the shale layer at extremely high pressures in order to release the trapped natural gas. It is a water-intensive practice. Typical projects use 1-3 million gallons of water for each well and 0.5 million pounds of sand. Large projects may require up to 5 million gallons of water.

The Texas Railroad Commission estimates that fracturing a vertical well in the Barnett Shale in Texas can use more than 1.2 million gallons of water, while fracturing a horizontal well can use more than 3.5 million gallons.[57] Moreover, the wells may be re-fractured several times, thus requiring additional water. Fracturing operations use an estimated 5 or more million gallons of water each day in the Barnett Shale, a smaller natural gas field in Texas.[58] Regarding the Marcellus Shale region, the USGS observed "many regional and local water management agencies are concerned about where such large volumes of water will be obtained, and what the possible consequences might be for local water supplies."[59]

Some of the injected fluids remain trapped underground, but the majority of the injected water—60% to 80%—returns to the surface as "flowback" after the frac treatment. It typically contains proppant (sand), chemical residue, and trace amounts of radioactive elements that naturally occur in many geologic formations.[60] USGS notes that because the quantity of fluid used is so large, the

---

[56] Daniel J. Soeder and William M. Kappel, *Water Resources and Natural Gas Production from the Marcellus Shale*, U.S. Geological Survey, U.S. Department of the Interior, Fact Sheet 2009-3032, May 2009, http://pubs.usgs.gov/fs/2009/3032/pdf/FS2009-3032.pdf. Hereafter, USGS Fact Sheet.

[57] Railroad Commission of Texas, Water Use in the Barnett Shale, July 30, 2008, http://www.rrc.state.tx.us/division/og/wateruse_barnettshale.html.

[58] Burnett, D.B. and Vavra, C.J., Desalination of Oil Field Brine—Texas A&M Produced Water Treatment. August 2006, http://www.pe.tamu.edu/gpri-new/home/BrineDesal/MembraneWkshpAug06/Burnett8-06.pdf.

[59] USGS Fact Sheet at 4.

[60] These particles, termed normally occurring radioactive materials (NORMS), can be brought to the surface on drilling equipment and in fluids. Subsurface formations may contain low levels of such materials as uranium and thorium and

(continued...)

---

additives in a three million gallon frac job would yield about 15,000 gallons of chemicals in the waste.[61] The well service company may temporarily retain the flowback and brine in open-air, lined retention ponds before reusing it (if possible), or disposing it. The drilling operator must reclaim the temporary storage pits when the drilling and fracturing operations end. In addition, the well operator must separate, treat, and dispose the natural brine co-produced with gas. As noted below, where feasible, the produced water may be disposed through underground injection. The oil and gas industry uses this disposal method in some western states and in Ohio.[62] The industry has not yet begun using it as a disposal alternative for gas production in eastern Marcellus Shale.

In the event that underground injection is not feasible in the area of the Marcellus Shale, the well service company may discharge the flowback to surface waters if the discharge does not violate a stream or lake's water quality standards. Standards established by states under Section 303 of the Clean Water Act (CWA) protect designated beneficial uses of surface waters, such as recreation or public water supply.[63]

If contaminants present in the flowback prevent discharge to surface water without further treatment, it is likely that the service company will have to transfer the wastewater off-site to an industrial treatment facility or a municipal sewage treatment plant that is capable of handling and processing the wastewater. In this case, the operator of the publicly owned treatment works (POTW) or industrial treatment facility would assume responsibility for treating the waste before discharging it into nearby receiving water in compliance with effluent limits contained in the facility's discharge permit.[64] The chemical frac additives returned in flowback and the produced brine could cause problems for POTWs. Contaminants in industrial process wastewaters can kill off the biota essential to a POTW's operation. If contaminants pass through the POTW without adequate treatment, the discharge could violate the facility's discharge permit and could cause a violation of water quality standards. A standard POTW's effective treatment of flowback and brine is uncertain. It could require upgrading, but the cost of such an upgrade is also uncertain.

In the fall of 2008, water samples from the mid-Monongahela River valley of Pennsylvania showed high levels of total dissolved solids (TDS), which indicate salinity. Although the TDS was determined to pose little threat to health or safety, preliminary analysis suggested that the principal source was large truck deliveries of wastewater from gas well drilling sites in the Marcellus Shale to POTWs discharging, directly or indirectly, into the Monongahela River. In October 2008, state officials ordered nine sewage treatment plants to reduce their volumes of gas well drilling water, which contains high concentrations of TDS. Subsequent analysis concluded

---

(...continued)

their daughter products, radium 226 and radium 228. On gamma-ray logs, shales can be differentiated from other rocks such as clean sandstones and limestones because shales have higher concentrations of potassium-40-bearing minerals. See Commonwealth of Pennsylvania Department of Conservation and Natural Resources, *Pennsylvania Geology*, Vol. 38, No. 1, p. 5, Spring 2008, http://www.dcnr.state.pa.us/topogeo/pub/pageolmag/pdfs/v38n1.pdf.

[61] USGS Fact Sheet at 4.

[62] In the Barnett Shale area, most of the water is reinjected for disposal.

[63] 33 U.S. C. § 1313.

[64] Under CWA Section 301, it is illegal to discharge pollutants into the nation's waters except in compliance with substantive and procedural provisions of the law, which include obtaining a discharge permit. 33 U.S.C. § 1311.

that discharge from abandoned mines was more responsible for the high TDS than drilling wastewater discharges from municipal wastewater treatment plants.[65] However, state officials remain concerned about the projected need for treatment of wastewater (both initial flowback water from fracturing and longer term production brines) from gas well development—projected to be as much as 20 million gallons per day in 2011—and the capacity of the state's surface waters to assimilate associated wastewaters. In April 2009, the Pennsylvania Department of Environmental Protection issued a permitting strategy document for gas development wastewaters, requiring that any discharges will be subject to the most stringent treatment or water quality standards needed to protect aquatic life in the state's streams. Their goal is prohibiting new sources of high-TDS wastewaters from discharging into Pennsylvania's waters by January 1, 2011.[66]

Brine storage and transport are major issues in developing the Barnett Shale in Texas, and are likely to be key issues in development of the Marcellus Shale, as well. Currently, permitted treatment facilities capable of treating such wastes are not adequate. In Pennsylvania, there are five facilities designed to treat the type of industrial wastewater that is involved in producing shale gas. Most of the well sites are located in northeast Pennsylvania, while the closest treatment facilities are nearly 250 miles away.[67]

West Virginia, too, recognizes that wastewater disposal is "perhaps the greatest challenge regarding these operations."[68] State officials say that underground injection control (see discussion below) may be the best option for wastewater disposal, but the state has only permitted two commercial underground injection control (UIC) wells. The state has no centralized commercial treatment facilities available, and state officials are cautious about the capability of POTWs to handle the flow and quality of waste that they might receive.[69] The West Virginia Department of Environmental Protection has proposed both changes to the state's oil and gas drilling rules (which the state legislature must approve) and an industry guidance document to assist operators in planning for the water issues associated with drilling and operating these wells. However, local groups have criticized the proposed rules and draft non-binding guidance for failing to address disposal of wastewater, disclosure of chemicals used in hydraulic fracturing, and where the additional quantities of water required for drilling will come from.[70]

---

[65] "Minimal Impact on Total Dissolved Solids Found in Monongahela River Last Fall," Natural Gas, Tapping Pennsylvania's Potential, May 20, 2009, http://www.pamarcellus.com/news.php.

[66] Marcellus Shale Wastewater Partnership, "Permitting Strategy for High Total Dissolved Solids (TDS) Wastewater Discharges," April 11, 2009, http://www.depweb.state.pa.us/watersupply/cwp/view.asp?a=1260&Q=545730& watersupplyNav=|.

[67] Legere, Laura, "How to handle wastewater big challenge in gas drilling," The Citizens' Voice, August 25, 2008.

[68] State of West Virginia, Department of Environmental Protection, Office of Oil and Gas, "Industry Guidance, Gas Well Drilling/Completion, Large Water Volume Fracture Treatments (Draft)," March 13, 2009, p. 3, http://www.wvsoro.org/curent_events/marcellus/Marcellus_Guidance_Draft.pdf.

[69] Ibid. at 3-4. Reportedly, one company with wells in the Marcellus shale in West Virginia has its hydraulic fracturing wastewater trucked to an out-of-state commercial facility that treats the water and then injects in into depleted oil and gas reservoirs. (Kasey, Pam, *New Drilling Efforts Raise Questions*, The State Journal, August 14, 2008.)

[70] West Virginia Surface Owners' Rights Organization, "Proposed Changes to Oil & Gas Rules, Marcellus Guidance Document," http://www.wvsoro.org/curent_events/.

One potential solution to off-site disposal may be on-site treatment and reuse; that is, treating and reusing flowback and produced water on-site. Some companies are reportedly considering on-site treatment options such as advanced oxidation and membrane filtration processes. On-site treatment technologies may be capable of recovering 70%-80% of the initial water to potable water standards, thus making the water immediately available for reuse. The remaining 20%-30% is very brackish and considered brine water. A portion may be further recoverable as process water, but not to achieve potable water standards. In other cases, companies send the brine water off-site for treatment and disposal. The economics of any such options are critical, and site factors such as available power and final water quality are often the determinant in treatment selection.

Other Surface Water Quality Issues

Another potential source of water pollution from oil and gas drilling sites is runoff that occurs after a rainstorm. The storm water runoff can transport sediment to nearby surface water bodies. Provisions of the CWA generally regulate storm water discharges from industrial and municipal facilities by requiring implementation of pollution prevention plans and, in some cases, remediation or treatment of runoff.[71] Industries that manufacture, process, or store raw materials and that collect or convey storm water associated with those activities are subject to the act's requirements. Furthermore, fracturing fluid chemicals and wastewater can leak or spill from injection wells, flow lines, trucks, tanks or holding pits and thus may contaminate soil, air, and water resources.

However, the act specifically exempts the oil and gas industry from these storm water management regulatory provisions. CWA Section 402(l)(2) exempts mining operations or oil and gas exploration, production, processing, or treatment operations or transmission facilities from federal storm water regulations, and Section 502(24) extends the exemption to construction activities, as well.[72] Thus, federal law contains no requirements to minimize uncontaminated sediment pollution from the construction or operation of oil and gas operations. However, the federal exemption does not hinder states from requiring erosion and sedimentation controls at well sites, under authority of non-federal law. Pennsylvania, for example, requires well drill operators to obtain a permit for implementation of erosion and sedimentation controls, including storm water management, if the site disturbance area is more than five acres in size. If the site is less than five acres, a plan for erosion and sediment control is required. Storm water requirements are part of this permit.[73] New York has similar requirements for erosion and sedimentation controls at well sites, regardless of site area. The Delaware River Basin Commission, which has jurisdiction over water quality in a portion of the area underlain by the Marcellus Shale (see section on State Water Quality Laws below) also has similar requirements regardless of site area.

---

[71] Clean Water Act section 402(p); 33 U.S.C. § 1342(p).

[72] 33 U.S.C. § 1342(l)(2); 33 U.S.C. §1362(24).

[73] The Pennsylvania permit is called an Earth Disturbance Permit (ESCGP-1).

Groundwater Quality Protection

Because development of the Marcellus Shale is dependent on the use of hydraulic fracturing, some fear it could potentially contaminate underground aquifers that provide water supplies to private wells and public water systems. The Safe Drinking Water Act in 2005 broadly excluded the underground injection of fluids used in hydraulic fracturing for oil and gas development.[74] However, the SDWA does not preempt states from imposing their own laws and regulations, and the states have long been responsible for protecting groundwater resources during oil and gas production activities.[75] For example, in New York, the Department of Environmental Conservation (DEC) has authority over oil and gas development in the state, including oversight of hydraulic fracturing activities to ensure protection of groundwater resources. Although federal laws do not regulate the injection of hydraulic fracturing fluids, states such as Pennsylvania and New York do require submission of information on hydraulic fracturing fluid composition prior to issuing a well permit. Moreover, other injection wells related to oil and gas development may be subject to federal requirements.

Safe Drinking Water Act Authority

The underground injection control provisions of the SDWA require the Environmental Protection Agency (EPA) to regulate the underground injection of fluids (including solids, liquids, and gases) to protect underground sources of drinking water. UIC program regulations specify sitting, construction, operation, closure, financial responsibility, and other requirements for owners and operators of injection wells.[76]

West Virginia, Ohio, and Texas are among the states that have assumed primacy and have lead implementation and enforcement authority for the UIC program, including primacy for injection wells associated with oil and gas development. EPA implements the programs directly for New York and Pennsylvania.[77] Most states, including Ohio and West Virginia, have received primacy for Class II oil and gas wells under Section 1425.

The Safe Drinking Water Act specifies that the UIC regulations may not interfere with the underground injection of brine or other fluids brought to the surface in connection with oil and gas production or any underground injection for the secondary or tertiary recovery of oil or natural gas

---

[74] EPA retains the authority to regulate the use of diesel fuel for the purpose of hydraulic fracturing if the agency considers such regulation necessary to protect underground sources of drinking water.

[75] SDWA § 1414(e); 42 U.S.C. § 300g-3(e).

[76] Application, construction, operating, monitoring, and reporting requirements for Class II wells are found in 40 CFR 144 and 146.

[77] To receive primacy, a state, territory, or Indian tribe must demonstrate to EPA that its UIC program is at least as stringent as the federal standards; the state, territory, or tribal UIC requirements may be more stringent than the federal requirements. For Class II wells, states must demonstrate that their programs are effective in preventing pollution of underground sources of drinking water (USDWs). Requirements for state UIC programs are established in 40 CFR §§ 144-147.

unless such requirements are essential to assure that underground sources of drinking water will not be endangered by such injections.[78]

Additionally, the Energy Policy Act of 2005 amended SDWA UIC provisions to specify further that the definition of "underground injection" excludes the injection of fluids or propping agents (other than diesel fuels) used in hydraulic fracturing operations related to oil, gas, or geothermal production activities.[79]

The key statutory provisions are:[80]

- SDWA Section 1421 directs EPA to promulgate regulations for state underground injection control (UIC) programs, and mandates that the regulations contain minimum requirements for programs to prevent underground injection that endangers drinking water sources.[81]

- Section 1422 authorizes EPA to delegate primary enforcement authority (primacy) for UIC programs to the states, provided that state programs prohibit any underground injection that is not authorized by a state permit.[82]

- Section 1425 provides separate authority for states to attain primacy specifically for oil and gas (i.e., Class II) wells. The provision allows states to demonstrate that their existing programs for oil and gas wells are effective in preventing endangerment of underground sources of drinking water, providing an alternative to meeting many of the detailed requirements promulgated to implement the UIC program under Section 1421.

- Section 1431 grants the EPA Administrator emergency powers to issue orders and commence civil action to protect public water systems or underground sources of drinking water. The Administrator may take action when (1) a contaminant present in or likely to enter a public drinking water supply system or underground drinking water source poses a substantial threat to public health, and (2) state or local officials have not taken adequate action.[83]

---

[78] SDWA §1421(b)(2)

[79] P.L. 109-58, Section 322, amended SDWA section 1421(d).

[80] SDWA §§ 1421 - 1426; 42 U.S.C. §§ 300h - 300h-5. The Safe Drinking Water Act of 1974 (P.L. 93-523) authorized the UIC program at EPA.

[81] Section 1421(d)(2) states:

> underground injection endangers drinking water sources if such injection may result in the presence in underground water which supplies or can reasonably be expected to supply any public water system of any contaminant, and if the presence of such contaminant may result in such system's not complying with any national primary drinking water regulation or may otherwise adversely affect the health of persons.

[82] P.L. 93-523, SDWA §1421 (42 U.S.C. § 300h).

[83] 42 U.S.C. § 300i.

Underground Injection of Waste Fluids

As noted, nearly all of the water injected for hydraulic fracturing must come back out of the well for gas to flow out of the shale. A key issue related to Marcellus Shale gas production is safely disposing large quantities of potentially contaminated fluids recovered from the gas wells.

EPA generally categorizes injection wells associated with oil and gas production as Class II wells under its UIC regulatory program. These are wells used to inject brines and other waste fluids associated with oil and natural gas production.[84] Given the expense of treating and transporting large volumes of wastewater for disposal, it is possible that the production of gas from the Marcellus Shale will increasingly involve the use of injection wells to dispose of poor-quality formation water, flowback water resulting from hydraulic fracturing, and other waste fluids associated with gas production.

EPA reports that most of the fluid injected into Class II wells has been brine brought to the surface in producing oil and gas. This brine, a naturally occurring formation fluid, is often very saline and may contain toxic metals and naturally occurring radioactive substances. According to EPA, the brine "can be very damaging to the environment and public health if it is discharged to surface water or the land surface."[85] To prevent contamination of land and surface water, Class II wells provide a means for disposing brines by re-injecting them back into their source formation or into similar formations. Injection wells also serve as disposal means for residual water from drilling and hydraulic fracturing operations. As states have adopted rules to prevent the disposal of saline water to surface water and soil, injection has become the preferred way to dispose of this waste fluid, where the local geology permits.[86]

In New York and Pennsylvania, both EPA and the state environmental agency must issue permits if the disposal method for fracturing wastewater is deep well injection. Pennsylvania law provides that "a well operator who affects a public or private water supply by pollution or diminution shall restore or replace the affected supply with an alternate source of water adequate in quantity and quality for the purposes served by the supply."[87] Additionally, it requires a permit application for a disposal well or enhanced recovery well to include an erosion and sedimentation plan for the well site.[88]

As of now, it is unknown how much water the gas wells in the Marcellus Shale formation will produce. The amount of water produced could vary across the region. Because shale gas formations generally are impermeable, they typically produce much less water than traditional oil

---

[84] Other Class II wells include oil and natural gas storage wells and enhanced oil and gas recovery wells.

[85] U.S. Environmental Protection Agency, Underground Injection Control Program. Oil and Gas Injection Wells: Class II, http://www.epa.gov/safewater/uic/wells_class2.html.

[86] The largest subclass of Class II wells are enhanced recovery wells. These wells are used to inject various substances (including brine, water, steam, polymers, and carbon dioxide) into hydrocarbon-bearing formations to recover primarily oil, but also natural gas, that remains in previously produced areas. Class II enhanced recovery wells are regulated under the UIC program, except for their use in hydraulic fracturing operations.

[87] 25 PA Code § 78.51. Protection of water supplies.

[88] 25 PA Code § 78.18. Disposal and enhanced recovery wells.

and gas fields or coalfields. The impermeability of the shale also indicates that reinjection of wastewater from fracturing into the shale formation may not be feasible in many locations. Consequently, it is uncertain whether Class II disposal wells will find wide use in the Marcellus Shale formation. Currently, only four injection wells operate for this purpose in Pennsylvania.

Wastewater injection into the permeable Cambrian sandstones that lie beneath the Marcellus Shale appears feasible. The Cambrian Mt. Simon Sandstone, considered an ideal geologic unit in Ohio for disposal and long-term storage of liquid wastes, is relatively deep, and underlain and overlain by impervious confining layers that prevent migration of injected fluids.[89]

Although underground injection of wastewater may not be practical or economic in all areas across the Marcellus region because of the lack of suitable injection zones, the cost and environmental concerns associated with surface disposal may make Class II injection wells the preferred disposal option for flowback and other wastewater from hydraulic fracturing operations where feasible. This appears to be the trend in other shale areas. In the Fayetteville Shale in Arkansas, trucks have typically collected wastewater and hauled it to disposal wells distant from the producing areas. However, with more intense shale development, the high cost of transporting, treating, and disposing water offsite, injection well use has increased.[90] In the Barnett Shale in Texas, flowback water has been primarily disposed in Class II injection wells.

Both Class II injection and municipal and industrial water treatment facilities are under consideration for the Marcellus Shale, and more than 60 permit applications for such wells are pending in New York for Marcellus Shale development.[91] One firm active in Marcellus Shale development has been disposing of flowback water and produced water using three UIC disposal wells and two commercial water treatment facilities, but reportedly plans to use only disposal wells in the future. Based on leases already held, the firm plans to drill between 13,500 and 17,000 gas wells.[92]

Technical and practical questions regarding the development of the Marcellus Shale remain unanswered. USGS researchers have noted that while drilling and hydraulic fracturing technologies have improved over the past several decades, "the knowledge of how this extraction might affect water resources has not kept pace."[93] Consequently, environmental regulators and gas developers face new challenges and some uncertainties as the Marcellus Shale is developed.

---

[89] Ohio Department of Natural Resources, Division of Geological Survey, *The Geology of Ohio—The Cambrian*, GeoFacts No. 20, May 1998.

[90] University of Arkansas and Argonne National Lab, Reducing the Environmental Impact of Natural Gas Development, http://lingo.cast.uark.edu/LINGOPUBLIC/.

[91] J. Daniel Arthur, et al, Evaluating the Environmental Implications of Hydraulic Fracturing in Shale Gas Reservoirs, 2008, available at http://www.all-llc.com.

[92] StatoilHydro, Frequently Asked Questions: *StatoilHydro's Acquisition of 32.5% of Chesapeake Interest in the Marcellus Shale*, 2008. http://www.statoilhydro.com/en/NewsAndMedia/News/2008/Downloads/Frequently%20asked%20questions.pdf

[93] USGS Fact Sheet at 5.

State Water Quality Laws

State laws addressing the quality of surface water and groundwater also appear to apply to Marcellus Shale development. For example, in New York various aspects of such development would require a permit under the State Pollutant Discharge Elimination System (SPDES).[94] SPDES is an "approved," rather than delegated, version of the federal National Pollutant Discharge Elimination System (NPDES) permit program because, while the federal NPDES covers only discharges to surface water, SPDES covers discharges to groundwater also. The SPDES permit requirement could apply to hydraulic fracturing by meeting four conditions:

1. Most importantly, the state must determine that injection will not degrade groundwater;[95]

2. A wastewater treatment plant would likely dispose of the fluids produced from the well, in which case the plant's SPDES permit would apply;

3. SPDES permits would also cover treatment facilities built specially for disposing of flowback water, if there would be discharges into a water body; and

4. Applicable state water-quality standards would control, in part, the permit's discharge limits.[96]

New York State's Environmental Quality Review Act (SEQRA) is also relevant.[97] As with its federal counterpart, the National Environmental Policy Act, a requirement of an environmental impact statement preparation lies at the heart of the statute.[98] New York has been evaluating the environmental impact of the drilling and hydraulic fracturing activities for more than 15 years through a Generic Environmental Impact Statement (GEIS) that sets parameters that apply statewide for SEQRA review of gas well permitting. In February 2009, the state's Bureau of Oil and Gas Regulation, in the Department of Environmental Conservation, released the final scoping document under SEQRA for a Supplemental Generic Environmental Impact Statement (SGEIS) on developing the Marcellus and other gas shale regions in the state using high-volume hydraulic fracturing. On September 30, New York DEC released for comment the draft SGEIS which proposes additional parameters for SEQRA review and focuses on water supply protection and wastewater management as major issues. Until New York finalizes the supplemental GEIS, the state will only accept, but not process, permit applications for gas wells involving horizontal drilling and high-volume hydraulic fracturing.[99]

---

[94] N.Y. Envtl. Cons. Law § 17-0505.

[95] N.Y. Code of Rules and Regulations (Conservation) § 750-1.5(a)(6).

[96] N.Y. Envtl. Cons. Law § 17-0501.

[97] N.Y. Envtl. Cons. Law §§ 8-0101 – 8-0117.

[98] Id. at § 8-0109.

[99] In the announcement of the draft supplemental GEIS, the NY Department of Environmental Conservation noted that,

> While the process of preparing the Supplemental GEIS is ongoing, any entity that applies for a drilling permit for horizontal drilling in the Marcellus Shale and opts to proceed with its permit application will be required to undertake an individual, site-specific environmental review. That review must take into account the same issues being considered in the Supplemental GEIS process

(continued...)

As another example, West Virginia's NPDES permit program would apply to wastewater treatment plants to which flowback from Marcellus Shale production sites was taken and to treatment facilities built specially for the frac water that discharge into a water body.[100] Applicable state water-quality standards would control the permit's discharge limits, in part.[101] However, this program applies to surface water only, not groundwater, and the state's Groundwater Protection Act exempts "groundwater within geologic formations which are site specific to ... [t]he production ... of ... natural gas.... "[102] The state's underground injection control program would regulate frac water re-injected at a second or subsequent production site.[103]

In addition to state water-quality laws, the interstate Delaware River Basin Commission (36% of whose jurisdictional land area in Pennsylvania and New York overlies the Marcellus Shale formation) would also impose water quality requirements.[104] The Commission's water quality (and other) requirements are legally separate from those of the affected states—that is, obtaining state approval does not excuse an applicant from seeking Commission approval—though in some cases the two requirements may be substantively identical.

Another interstate-compact-created commission within the Marcellus region, the Susquehanna River Basin Commission, regulates only water quantity, not water quality.

State Water Supply Management

Gas producers must arrange to procure the large volumes of water required for hydraulic fracturing in advance of their drilling and development activity. Generally, water rights and water supply regulation differ among the states. Depending on individual state resources and historic development, states may use one of two water rights doctrines, riparian or prior appropriation, or a hybrid of the two. Under the riparian doctrine, a person who owns land that borders a watercourse has the right to make reasonable use of the water on that land.[105] Traditionally, the

---

(...continued)

and must be consistent with the requirements of the State Environmental Quality Review Act and the state Environmental Conservation Law.

See http://www.dec.ny.gov/energy/46288.html.

[100] W. Va. Code Ann. § 22-11-4(a)(16). See regulations at W. Va. Code of State Rules tit. 47, ser. 10.

[101] W. Va. Code of State Rules tit. 47, ser. 2.

[102] W. Va. Code Ann. § 22-12-5(i).

[103] W. Va. Code Ann. § 22-11-8(b)(7). See regulations at W. Va. Code of State Rules tit. 47, ser. 13.

[104] The compact creating the Delaware River Basin Commission was ratified by Congress: P.L. 87-328, 75 Stat. 688. Section 3.8 of the Compact states: "No project having a substantial effect on the water resources of the basin shall hereafter be undertaken by any person, corporation, or government authority unless it shall have been first submitted to and approved by the commission.... " Section 2.3.5 B of the Delaware River Basin Comm'n Administrative Manual (Rules of Practice and Procedure) lists 18 types of projects that must be submitted to the Commission, including withdrawal of groundwater and discharge of pollutants into surface or ground waters of the Basin. Codified at 18 C.F.R. § 401.35(b).

[105] *See generally* A. Dan Tarlock, Law of Water Rights and Resources, ch. 3 "Common Law of Riparian Rights."

---

only limit to users under the riparian system is the requirement of reasonableness in comparison to other users. Under the prior appropriation doctrine, a person who diverts water from a watercourse (regardless of his location relative thereto) and makes reasonable and beneficial use of the water may acquire a right to use of the water.[106] The prior appropriation system limits users to the quantified amount of water the user secured under a state permitting process with a priority based on the date the state conferred the water right. Because of this priority system, the phrase "first in time, first in right" has sometimes substituted for appropriative rights. Some states have implemented a dual system of water rights, assigning rights under both doctrines.

Generally, states east of the Mississippi River follow a riparian doctrine of water rights, while western states follow the appropriation doctrine.[107] The system of water rights allocation in a particular state with shale resources may affect the development process, particularly in times when shortages in water supply affect the area of shale development. In areas where the Marcellus Shale is located, which are generally riparian states, water rights may not be as big a concern as in other areas of the country with shale development, such as the Barnett Shale in Texas. That is, even in times of shortage, shale development may be able to continue in the Marcellus Shale region because riparian users reduce water usage proportionally and may still receive enough for supply requirements of the development process. On the other hand, appropriative rights users in the Barnett Shale region may not be able to fill their water rights at all if other senior rights take all the water, and thus would have to postpone development. In addition, interstate compacts may apply and affect water supply for shale development processes. In the case of Marcellus Shale development, several interstate compacts are relevant, as discussed below.

New York's SPDES permit program (discussed above) governs water quality only, not water quantity. With a limited exception for pumping water on Long Island,[108] there is no proactive regulatory scheme in New York for extracting water from streams, lakes, groundwater, etc. In the case of drawing water from a public drinking water supplier, however, the state does have limited authority to make sure that the public water supplier stays within its permit terms. Otherwise, however, the state can only *respond* to water flow problems—e.g., if a fish kill occurs, it can prosecute the responsible entity for violating the flow standard that is a component of the state's water quality standards.[109] There is no requirement to notify the state in advance of a water extraction.

---

[106] *See generally id.* at ch. 5, "Prior Appropriation Doctrine."

[107] The distinction between these doctrines arises primarily from the historic availability of water geographically. In the generally wetter, eastern riparian states, water users share the water resources because water availability historically did not pose a problem to settlement and development. In the drier, western states that experience regular water shortages, the prior appropriation system provides a definitive hierarchy that allows users to acquire well-defined rights to water as a limited resource that requires planning to avoid scarcity.

[108] N.Y. Code of Rules and Regulations (Conservation) § 602.1.

[109] N.Y. Code of Rules and Regulations (Conservation) § 703.2. For certain classes of water bodies, the flow standard prohibits any "alteration that will impair the waters for their best usages."

West Virginia passed the Water Resources Protection and Management Act in 2003.[110] It requires users of water resources whose withdrawals exceed 750,000 gallons in any given month for one facility to register with the Division of Water and Waste Management in the Department of Environmental Protection.[111] To protect both ground and surface waters, the state proposes to require operators to provide information about the sources of withdrawals, anticipated volumes, and the time of year of withdrawals prior to start-up.[112] State officials believe it is likely that some oil and gas industry operations in the Marcellus Shale region will exceed this threshold and will be required to submit withdrawal information.[113] The goal is to ensure that water withdrawal from ground or surface waters does not exceed volumes beyond what the waters can sustain.[114]

Texas is another relevant example, because of similarities between the Barnett Shale there and the Marcellus Shale. Texas has codified the public trust doctrine regarding ownership of state water resources. That is, water in any of the various waterbodies—including rivers, streams, lakes, etc.—within the state is the property of the state of Texas.[115] Individuals or entities may divert the state's waters for their own use only after acquiring a permit (water right) from the state through its Commission on Environmental Quality.[116] Texas does provide for the possibility of temporary water permits for a period of up to three years, if a temporary permit would not adversely affect senior rights.[117]

Other states apply surface and groundwater regulations similarly, and gas producers using fresh water for drilling and development must comply with state and local administration of water rights.

As for interstate constraints in the Marcellus Shale region and vicinity, the Delaware River Basin Commission[118] and the Susquehanna River Basin Commission[119] impose limits on the quantity of water extracted. In addition, the Great Lakes-St. Lawrence River Basin Water Resources Compact[120] prohibits inter-basin transfers of water.

---

[110] W. Va. Code Ann. § 22-26.

[111] *Id.* at § 22-26-3.

[112] *See id.*

[113] *See* State of West Virginia, Department of Environmental Protection, Office of Oil and Gas, "Industry Guidance, Gas Well Drilling/Completion, Large Water Volume Fracture Treatments (Draft)," March 13, 2009, pp, 1-2, http://www.wvsoro.org/curent_events/marcellus/Marcellus_Guidance_Draft.pdf.

[114] *Id.*

[115] Tex. Water Code § 11.021.

[116] *See* Tex. Water Code § 11.022.

[117] Tex. Water Code § 11.138.

[118] Congress ratified the compact creating the Delaware River Basin Commission in Pub. Law 87-328, 75 Stat. 688.

[119] Congress ratified the compact creating the Susquehanna River Basin Commission in Pub. Law 91-575, 84 Stat. 1509.

[120] For the text of the compact, see http://www.cglg.org/projects/water/docs/12-13-05/Great_Lakes-St_Lawrence_River_Basin_Water_Resources_Compact.pdf. Congress ratified the compact in P.L. 110-342, 122 Stat. 3739.

Congressional Interest

Recently proposed legislation would affect natural gas projects in the Marcellus Shale directly and indirectly. The 111th Congress introduced bills to amend the Safe Drinking Water Act to define hydraulic fracture as underground injection for regulatory purposes. Bills to revise the Commodity Exchange Act would place limits on the use of futures markets to hedge the risks associated with natural gas development projects. The recently enacted mandate to increase the use of renewable fuels (ethanol) has an indirect link, as fertilizer produced from natural gas is essential to producing corn — the primary feedstock for ethanol. Recently proposed low-carbon fuel standards may overlook natural gas as a substitute transportation fuel.

Current and planned projects to develop Marcellus Shale gas are apparent across the six-state region that overlies the resource (see **Figure 5**). For example, gas producers have reportedly planned over 2,000 gas wells just in West Virginia, and the state's Oil and Gas Commission estimates that, based on current information, there could be a well on every 40 acres in the state. Throughout the region, this activity is likely to put increasing demands on regulatory agencies— especially state agencies[121]—for necessary licensing, permitting, inspections, and enforcement. As noted by USGS, because of questions related to water supply and wastewater disposal, many state agencies have been cautious about granting permits, and some states have placed *de facto* moratoria on drilling until these issues are resolved.[122] New York, for example, is accepting but not processing permits until it completes a Supplemental Generic Environmental Impact Statement that will impose new environmental review requirements for permits for gas well development using high-volume hydraulic fracturing. The success of planned development activities could depend, in part, on the capacity of regulatory agencies to provide the administrative resources that supporting such plans would require.

The Energy Policy Act of 2005 (P.L. 109-58) in Section 322 (Hydraulic Fracturing) amended the Safe Drinking Water Act (42 U.S.C. 300h(d)(1)) to exclude the fluids or propping agents (other than diesel fuels) used in hydraulic fracturing operations related to oil, gas, or geothermal production activities from the definition of the term "underground injection." Several pending bills address the current exemption of hydraulic fracturing under SDWA:

- H.R. 2300 (introduced May 7, 2009, as the American Energy Innovation Act) would express the sense of Congress that the Safe Drinking Water Act was never intended to regulate natural gas and oil well construction stimulation and that the amendment of SDWA by the Energy Policy Act of 2005 to clarify that the SDWA was not intended to regulate the use of hydraulic fracturing should be maintained.

- H.R. 2766/S. 1215 (introduced June 9, 2009, as the "Fracturing Responsibility and Awareness of Chemicals Act") would amend the SDWA definition of 'underground injection' to include the underground injection of fluids or

---

[121] As discussed above, states have primary responsibility for many regulatory programs, several interstate commissions also are involved, as are federal agencies such as EPA that implements the UIC program and issues UIC permits in New York and Pennsylvania.

[122] USGS Fact Sheet at 5.

propping agents used for hydraulic fracturing operations related to oil and gas production activities. The bills would require public disclosure of the chemical constituents (but not the proprietary chemical formulas) used in the fracturing process. Disclosure of a propriety formula to the state, EPA Administrator, or treating physician or nurse would be required in the case of a medical emergency.

- The conference report for the Department of the Interior, Environment, and Related Agencies Appropriations Act, 2010 (H.R. 2996, H.Rept. 111-216) includes a provision urging the EPA, in consultation with appropriate federal, state and interstate agencies, to carry out a study on the relationship between hydraulic fracturing and drinking water.

Many industries make use of various "hedging" strategies to minimize the risk of commodity price increases. A simple hedge involves buying "futures" contracts to lock in prices. For gas exploration and development companies, hedges in effect guarantee the amount of revenue that companies will receive on a future production, thus giving them some financial stability. Two current proposed bills would limit speculation on future commodity prices. Some would argue, however, that in a time of low natural gas prices and scarce capital for financing new energy projects, restrictions on hedging could adversely these companies.

- H.R. 977 (introduced February 11, 2009, as the "Derivatives Markets Transparency and Accountability Act of 2009") would amend the Commodity Exchange Act to bring greater transparency and accountability to commodity markets. The Commodity Futures Trading Commission would establish limits on the amount of positions that a person may hold for future delivery contracts, options on such contracts or on commodities traded.

- S. 447 (introduced February 13, 2009, as the "Prevent Excessive Speculation Act") would amend the Commodity Exchange Act to prevent excessive price speculation with respect to energy commodities.

Congressional concern for energy independence that grew out of the summer 2008 oil price spikes has engendered at least one highly publicized proposal to substitute natural gas for transportation fuel.[123] Among various bills introduced to reduce consumer dependence on fossil fuels, some would establish low-carbon fuel standards. Natural gas (CH_4) has a carbon to hydrogen ratio of 1:4, the lowest of all the fossil fuels, thus leading some to argue that it has a role in newly proposed standards.

- H.R. 1787 (introduced March 30, 2009, and cited as the "Low Carbon Fuel Standard of 2009") would amend the Clean Air Act to establish a low carbon fuel standard for transportation fuels. Based on a lifecycle greenhouse gas emission baseline established for all transportation fuels, transportation fuel providers would have to reduce the annual emissions per unit of energy by 5% after 2023, and 10% after 2030.

- S. 1095 (introduced May 20, 2009, and cited as "America's Low-Carbon Fuel Standard Act of 2009") would amend the Clean Air Act to convert renewable fuel standards into low-carbon fuel standards. Low-carbon fuel would be defined

---

[123] See PickensPlan, http://www.pickensplan.com/act/.

as a transportation fuel that has lifecycle greenhouse gas emissions, equal on an annual average basis to a defined percentage less than baseline lifecycle greenhouse gas emissions. Starting at 20% in 2015, the percentage would increase to 42.5% after 2031. Regulations would insure that increasing volumes of low-carbon fuel would be sold in the United States as transportation fuel; beginning with 10% in 2015, and 32.5% by 2030.

U.S. fertilizer production has a close link to energy availability, particularly natural gas.[124] Natural gas is the key ingredient in producing anhydrous ammonia, used directly as a nitrogen fertilizer, and used as a basic building block for producing most other forms of nitrogen fertilizers (e.g., urea, ammonium nitrate, and nitrogen solutions). Natural gas also serves as a process gas in the manufacture of other nitrogenous fertilizers from anhydrous ammonia. As a result, natural gas accounts for 75% to 90% of costs of production for nitrogen fertilizers. Because fertilizer prices are closely linked to natural gas prices, higher gas prices encourage two potential responses: (1) lower fertilizer application rates on the current farm planting mix; or (2) the planting and production of crops that are less dependent on fertilizer. Among major U.S. field crops, corn uses the most fertilizer according to the U.S. Department of Agriculture's Economic Research Service.[125] Thus, ethanol production costs and thus renewable energy costs are likely to reflect natural gas availability and price. Under the Energy Independence and Security Act of 2007 (P.L. 110-140), gasoline sold in the United States must contain a minimum volume of renewable fuel. The Renewable Fuel Standard program that results from this requirement will increase the required volume of renewable fuel blended into gasoline from 9 billion gallons in 2008 to 36 billion gallons by 2022.

For Further Reading

For further reading on natural gas issues, refer to the following CRS reports:

- CRS Report R40872, *U.S. Fossil Fuel Resources: Terminology, Reporting, and Summary*, by Gene Whitney, Carl E. Behrens, and Carol Glover.

- CRS Report R40487, *Natural Gas Markets: An Overview of 2008*, by Robert Pirog.

- CRS Report R40645, *U.S. Offshore Oil and Gas Resources: Prospects and Processes*, by Marc Humphries, Robert Pirog, and Gene Whitney.

- CRS Report RL34741, *Drilling in the Great Lakes: Background and Issues*, coordinated by Pervaze A. Sheikh.

---

[124] For further information refer to CRS Report RL32677, *Energy Use in Agriculture: Background and Issues*, by Randy Schnepf.

[125] Wen-yuan Huang, William McBride, and Utpal Vasavada, *Recent Volatility in U.S. Fertilizer Prices*, USDA Economic Research Service, March 2009, http://www.ers.usda.gov/AmberWaves/March09/Features/FertilizerPrices.htm.

Appendix. Glossary

Bcf: Billion Cubic Feet; a gas measurement equal to 1,000,000,000 cubic feet. See also Mcf, Tcf, Quad.

Btu: British thermal unit; the amount of energy required to heat one pound of water by one degree Fahrenheit.

Frac: Hydraulic fracturing, as adopted by the petroleum industry.

Flowback: The fracture fluids that return to surface after a hydraulic fracture is completed.

Mcf: A natural gas measurement unit for one thousand cubic feet.

MMcf: A natural gas measurement unit for one million cubic feet.

NORM: Natural occurring radioactive material; includes naturally occurring uranium-235 and daughter products such as radium and radon.

Oil-equivalent gas (OEG): The volume of natural gas needed to generate the equivalent amount of heat as a barrel of crude oil. Approximately 6,000 cubic feet of natural gas is equivalent to one barrel of crude oil.

Slickwater: Water-based frac fluid.

Tcf: A natural gas measurement unit for one trillion cubic feet.

Thixotropy: The property of a gel to become fluid when disturbed (as by shaking).

Whipstock: A wedge-shaped piece of metal placed downhole to deflect the drill bit.

Author Contact Information

Anthony Andrews, Coordinator
Specialist in Energy and Energy Infrastructure
Policy
aandrews@crs.loc.gov, 7-6843

Peter Folger
Specialist in Energy and Natural Resources Policy
pfolger@crs.loc.gov, 7-1517

Marc Humphries
Analyst in Energy Policy
mhumphries@crs.loc.gov, 7-7264

Claudia Copeland
Specialist in Resources and Environmental Policy
ccopeland@crs.loc.gov, 7-7227

Mary Tiemann
Specialist in Environmental Policy
mtiemann@crs.loc.gov, 7-5937

Robert Meltz
Legislative Attorney
rmeltz@crs.loc.gov, 7-7891

Cynthia Brougher
Legislative Attorney
cbrougher@crs.loc.gov, 7-9121

Acknowledgments

Cynthia Brougher, Legislative Attorney

www.ingramcontent.com/pod-product-compliance
Lightning Source LLC
Chambersburg PA
CBHW081621170526
45166CB00009B/3055